Geoengineering
Not Chemtrails:

Investigation Into Humanities 6th Great Extinction

James W. Lee

Geoengineering not Chemtrails Book II
Title ID: 6443224
ISBN-13: 978-1535445979
Copyright ©2016 by James W. Lee
Geoengineering Not Chemtrails: Investigation Into Humanities 6th Great Extinction
All rights preserved. No part of this book may be reproduced in any form or by any electronic or mechanical means including information storage and retrieval systems without written permission form the author. Reviewers may quote brief passages to be printed in a magazine or newspaper.

Book designed and edited by Ellen Sklar Abbott

A Tribute

Everything flows, floats and moves. There is no state of equilibrium-there is no state of rest. —*Viktor Schauberger (1885–1958)*

Viktor Schauberger made an extraordinary contribution to knowledge of the natural world. He is celebrated for his huge respect for Nature.

He pioneered discoveries in the water sciences, in agricultural techniques, and in the energy domain-what enhances and what harms life. Mr. Schauberger showed that when the natural ecosystems are in balance and biodiversity rules, there is great creativity and evolution of higher and more complex life forms, but also order and stability.

When humans walked lightly on the Earth, we cooperated with Nature. While still part of her, we greatly abuse her and ourselves. Mr. Schauberger warned us eighty years ago that if we continued to go against Nature the Earth's ecosystems would become sick, the climate destructive, and human society would break down with extreme violence, greed, and pandemic illness.

Dedication

This book is dedicated to my son, Jaxson and all sentient children of the Earth. They deserve much better than the world we leave them today. Please forgive us adults who failed to halt Man's destruction of the environment on our watch.

Hopi Prophecy

Near the Day of Purification,
There will be Cobwebs spun
back and forth
in the
sky

Preface

People will do anything, no matter how absurd, in order to avoid facing their own souls. One does not become enlightened by imagining figures of light, but by making the darkness conscious. —*Carl Jung*

    The information provided in this book is a compilation of over a decade of research on climate weather modification also known as Geoengineering. It is a guided outline connecting vast geoengineering and weather manipulation practices by military, government, and private corporations who purposely misdirect and mislead. Geoengineering (GE) operation agencies demand that confidentiality agreements be signed so disseminating reality from fiction can be difficult.

    The material is compiled by diverse sources and may be perceived as very dark, dense, and deep. The term "chemtrail" is not recognized as a scientific term and those in power claim plausible denial and refuse accountability. Please take nothing in this book as fact or true. Ground yourself and allow time to process the many meanings behind the evil perpetrated on all of us

Factoid: Aerosol spraying over our heads had been going on for decades now without notice or alarm by the majority living on Earth at this time.

We must halt all geoengineering practices immediately and let are only Mother Earth heal before we no longer have that choice.

If geoengineering continues unabated, and the ozone layer is completely destroyed, it will be game over for us all. No ozone layer = no life on Earth.

It's that simple—

- Geoengineering is tearing apart the entire fabric of life on our planet.
- Geoengineering is poisoning our air, waters, and soils.
- Geoengineering is pushing increasingly erratic atmospheric processes resulting from human-propelled climate change past the Chaos threshold into unpredictable, self-reinforcing, cascading events.
- Geoengineering is disrupting the jet stream and all natural weather patterns, which in turn is fueling catastrophic climate feedback loops, the most dire of which are mass methane hydrate releases from the Arctic tundra and sea floor.
- Geoengineering is destroying the stratospheric solar radiation shielding, which protects all life on Earth.

CONTENTS

Tribute		I
Dedication		II
Preface		III
Chapter 1	Geoengineering Humanities 6th Great Extinction Event	1
Chapter 2	Geoengineering 101; Who, What, Where, How and Why	15
Chapter 3	Aerosol Spraying of Vaccines Ramps Up	21
Chapter 4	Get Your College Degree in Geoengineering	27
Chapter 5	The Great Arctic Melt Off	33
Chapter 6	The Trees are Dying; We're Next	45
Chapter 7	HAARP, LUCY, ALAMO, and A.N.G.E.L.S. Projects: Advanced Technology Acceleration to Early Extinction?	53
Chapter 8	Conclusion	71
Chapter 9	Summary	75
Epilogue		i
Appendix I		iii
Appendix II		xi
Bibliography, resources		xviii

Chapter 1 | Geoengineering Humanities 6th Great Extinction Event

If You Do Not Change Direction You May End Up Where You Are Heading   — *Lao Tzu*

Geoengineering–The artificial modification of Earths climate systems through two primary ideologies, Solar Radiation Management (SRM) and Carbon Dioxide Removal (CDR).

 Chemtrails is another name used as a psychological operation (PSYOP) because science does not recognize this term. This was done to purposely obfuscate and confuse while giving those who are practicing geoengineering to plausibly deny their covert and overt activities.
 Terms to use that government and science do recognize are: Climate Engineering, Strategic Aerosol Geoengineering (SAG), Solar Radiation Management (SRM), and even "Persistent Contrails," which is debunked as a possibility given today's jet engine technologies on most commercial and military aircraft flown today.

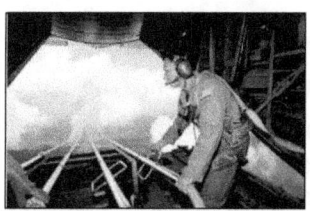

 Truth be told, not sold, the forests are dying, the oceans are dying, the Earth is dying, and even at this late hour the vast majority of first world societies remain in total ignorance, denial, and apathy manifested in many forms. The expanding list of extreme weather events and unfolding world die-offs are not discussed or disclosed by the power structure sold by a vast corporate media machine designed for mass distraction and disinformation.

Geoengineering Not Chemtrails:
Investigation Into Humanities 6th Great Extinction

Industrialized/militarized civilization is in its death throes, the pounding of Earth by covert climate engineering assault is a testimony to the near total loss of sanity in modern society today. Every breath we take is now toxic, air particulate pollution is taking immense toll on human health. The largest source of this atmospheric pollution, climate engineering, is systematically denied by the world's central banker controlled "experts" whose paid for opinions and biased research is blindly accepted as gospel truth by nearly everyone. Denial is not just a river in Egypt but a mass psychosis of the willful, yet social media correct, ignorant mass of "sheeple" among us all today who completely choose to ignore root and principle causes of weather and sentient life extinction gone wild around, and to, our world today.

> *"We have film footage of these tankers spraying at altitude, up close nozzles visible. We have close up photographs of these nozzles right behind the engines. At that point the argument ends. We have footage of the crime happening. There is no argument or dispute-it is absolutely going on. We have President Obama going on record saying that climate change is the greatest national security threat of all. Do you think that they would ask our permission before they do this? That is a very naive though."*
> —Dane Wiggington, www.geoengineeringwatch.org

Those who have the courage to speak truth-to-power about geoengineering operations above us almost daily face draconian consequences from a societal system that has all but abandoned any sense of reason or morality about our collective upcoming immortality because of climate chaos, made worse by Geoengineering, not to mention the legacy of the Baby "Doomers" who remain silent and sidelined while Earth is put in peril.

If you Google chemtrails or geoengineering and pick nearly any city in the world, you will see unaltered photographs taken in nearly every country you search, hiding in plain sight overhead. There are over 150 legal documents (US Patents) that evidence weather modifications, also known as geoengineering, that can be found in Appendix II of this book. Many of these patents cannot be employed or used unless applied through aerosol spraying.

Chapter 1 | Geoengineering Humanities 6th Great Extinction Event

We've Being Sprayed Like Bugs for Decades

Furthermore, what should be even more disturbing, is that we have all been unknowingly sprayed like insects without consent, since at least the mid 1940s. The US Military, with or without government approval, has used its citizens as lab rats for decades paid for with taxpayer dollars.

In 1955, "Operation Drop Kick," purposely released infected mosquitoes on poor African American populations in Georgia and Florida as part of the much larger Tuskegee Operation that lasted from 1932–1972, without prior knowledge. Hundreds of aerosol spraying operations were conducted over unsuspecting populations. In 1966 San Francisco residents were sprayed with *Serratia marcescens* bacteria to determine what biological weapons might simulate a germ-warfare attack. At the time, according to Rebecca Kreston for Discover Magazine, it was "one of the largest human experiments in history" and "one the largest offenses of the Nuremberg Code since its inception." The Serratia bacteria has been known to cause meningitis, arthritis, wound infections, and transfers through dialysis, blood transfusions, cauterization, and lumbar punctures.

> *"In short, chemtrails itself is a conspiracy theory"* —Puneet Killipara, *"How a Group of Conspiracy Theorists Could Derail the Debate Over Climate Policy,"* —January 22, 2015, Washington Post

Fast forward to today where retrofitted drone C141s and the like, regularly spray all skies, domestic and abroad, with chemicals containing aluminum, barium, strontium, lithium, formaldehyde, as well as nano-particulate matter called, Smart Dust, that we all have inhaled and ingested.

There is no such thing as organic food anymore. All soil, water, and air have nano-particulate matters from decades of almost daily spraying above us. The citation below is from an alleged whistle-blower using the name Locke, about spraying conducted in the spring of 2015 in Southern Oregon:

"For example, I could have warned thousands of people of the ongoing (as of January 2015) spraying to manufacture air stagnation in the Rogue and Umpqua [river] Valleys, and much of the Oregon Coast south of Florence. The artificially induced period of air stagnation is part of a larger experiment testing the efficacy of psychoactive chemical dispersal from high altitudes. Currently, spraying is most intense along the coast itself and above inland valleys. The stagnant air currents in the region allow for more direct application of psychoactive agents to test populations. I have little formal contact with the chemistry department, but to my knowledge, Lithium is the primary substance being dispersed in the aforementioned experiment.

Does this internet photo show the interior of a chemtrail plane?

Given the psychoactive and social nature of the ongoing tests, my department is playing a secondary role collecting water and soil samples. The Sociological Research Division however, has operatives throughout the region gathering a massive amount of data regarding the test population's behavioral traits like consumer habits, political engagement levels, and awareness of geoengineering programs. I would urge you to inform your friends and peers of this fact as avoiding observation by refusing to take surveys and engaging in similar actions that can disrupt this sordid research. The plumes disperse lithium and other psychoactive compounds. Areas targeted as of March 1, 2015, are inland valleys and coastal towns south of Florence. Over the next two months, additional spraying operations will be conducted to increase targeting areas in the Portland metropolitan area."

> Local and State air quality boards do not test for aerosol sprays made by man, hence they can honestly regularly report they find no evidence of geoengineering activities.

In both letters, he explains that he is "an employee of a weather data collection company and, by proxy, a subcontractor for the National Weather Service office in (central Oregon). In my work for the aforementioned institutions, I collected data, e.g., soil samples, that were used to direct spraying operations for the *last three years*."

According to a federal court ruling in April 2014, the US Army must inform all veterans of any potentially harmful health effects stemming from the secret medical and drug experiments conducted on the during the Cold War. According to a report by Courthouse News wire service, some 7,800 soldiers claimed to have been involved in these experiments. After

Chapter 1 | Geoengineering Humanities 6th Great Extinction Event

recruiting Nazi scientists to help assist in eugenics, mind control and rocket experiments through a program called "Project Paperclip," the Army and CIA administered between 250 and 400 kinds of drugs to the soldiers in an attempt to advance US ability to wage war. Among the many drugs used were Sarin, amphetamines, LSD, mustard gas, THC, incapacitating agents, and phosgene, a chemical weapon used in trenches during WWI. By administering these drugs and others, the military hoped to uncover new ways to control human behavior, pinpoint weaknesses, hypnotize, and increase an individual's resistance to torture.

The experiments began in the 1950s and continued until President Richard Nixon halted research into offensive chemical weapons in 1969 when the programs were pulled off the publics radar screens. Although soldiers signed consent forms agreeing to undertake the experiments, the soldiers argued in court they essentially had no other choice under training that directed them to follow orders. Veterans also argued that the forms violated international law and the Wilson Directive, which mandates voluntary consent as "essential."

Factoid: Soldier—known as "Sold to Die" by those in power, which is why they are required to wear "Dog Tags" as identification.

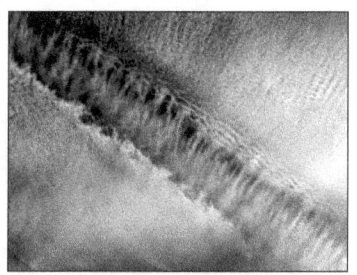

Fast forward to today and the very same technology perfected by spraying on the unobservant populace, introduced a new type of disease being labeled by the Alternative Media as "chemflu." Those of us who have been observing the sky painting going on over our heads for years have noticed of late a very appreciable increase in intensity and regularity of spraying as well as new types of aerosol sprays.

Longtime Geoengineering watchers have observed and noted many different kinds of aerosol spraying applications including scalar waves for mind control, with trails that drift lower spreading insipidly over the course of an hour or more as well as new types of spraying being labeled, "black chemtrails." For reasons unknown, when questioning FAA and other government regulatory agencies about spraying, they simply deny and discount with the use of propaganda campaigns, though over the decades evidence proves otherwise.

On my YouTube channel, *aplanetruth.info*, I recently posted a video documentary titled, *"Chemtrail Flu Going Viral Worldwide."* Soon after reports came in around the world of sudden increases in spraying in conjunction with flu like respiratory ailments, vomiting and stomach flu. This cannot be just another "coincidence theory."

Below is a list of some of the known chemicals found in the heavy metals used in geoengineering practices world-wide. *Ingredients are mixed in an aerosol cocktail* and contain barium chloride, aluminum oxide, synthetic polymers, bio-nano particles, and ethylene dibromide. An independent analysis of geoengineered fallout has conclusively identified many of these toxic chemicals:

Aluminum oxide particles
Arsenic
Bacilli and molds
Barium salts
Barium titanate
Cadmium
Calcium
Chromium
Desiccated human red blood cells
Ethylene bromide
Enterobacter cloaca
Enterobacteriaceae
Human white blood cells–A *(restrictor enzyme used in research labs to snip and combine DNA)*
Lead
Mercury
Methyl aluminum
Mold spores
Mycoplasma
Nano-aluminum-coated fiberglass
Nitrogen trifluoride (CHAFF)
Nickel

Compromised immunity
Disorientation
Difficulty paying attention and concentrating
Sinusitis
Skin discomfort and irritation
Joint pain
Muscle pain
Asthmatic (*breathing difficulties*)
Dizziness
Insomnia
Memory loss
Eye problems (*blurred or fuzzy vision*)
Nausea
Liver problems
Gallballder dysfunction
Tinnitus (*distant ringing in ears or high pitched sound after spraying*)
Neck pain
Scratchy throat
Allergy symptoms
Hay fever (*out of season*)
Flu-like symptoms
Susceptibility to colds
General weakness
Anxiety
Lightheaded or faint
Depression
Coughing
Sneezing
Shortness of breath
Vertigo
Anger, rage, frustration issues
MORGELLONS disease
   –List provided by *www.StopSprayingCalifornia.com*

Each of these symptoms is a normal occurrence in areas around the world where geoengineering practices have become a fact of life.

Polymer fibers
Pseudomonas aeruginosa
Pseudomonas florescens
Radioactive cesium
Radioactive thorium
Selenium
Serratia marcescens
Sharp titanium shards
Silver
Streptomyces
Strontium
Sub-Micron Particles (*containing live biological matter*)
Unidentified bacteria
Uranium
Yellow fungal mycotoxins

> *"If, however, millions of people are already on prescription pharmaceuticals to 'calm them down' and, in addition, are breathing poisoned air rife with mind-distorting chemicals, then how clearly is anyone able to think? How can anyone feel well and safe if the air we breathe is deliberately poisoned and is affecting our ability to think cogently? If is like Diogenes, the ancient Greek, searching for a truthful individual. No one seems to have the desire or courage, or authority to stop massive poisoning, because it is the secret plan of the elite insiders to deliberately destroy everything we once knew."*
> —Dr. Illya Sandra Perlingieri Global Research, May 12, 2010

What follows is a fairly exhaustive list of symptoms associated with aerosol spraying. Each symptom has been identified by various individuals who have clocked their occurrence with the onset of chemtrails being laid down over their homes or businesses. This list has been organized in a descending order with the most commonly experienced symptoms at the top.

Headache
Brain fog
Fatigue
Low energy

The preponderance of barium salt (barium chloride) and alumina (aluminum oxide), which are said to exist in the greatest concentration in *aerosol spraying*, is particularly alarming. Why? Because barium is well known to decrease and impair immune function. Aluminum oxide has its own set of problems, especially when inhaled in certain concentrations over prolonged periods of time. Aluminum, in any form is not very easy for the body to detoxify especially when it accumulates in the lungs.

The very last medical condition on the preceding list—*Morgellons disease*—is not a symptom, rather, it is a full blown disease process that usually takes years to manifest in its most serious form. In fact, there are websites dedicated exclusively to the research and treatment of *Morgellons disease*.

Cliff Carnicom (*www.carnicominstitute.com*) has been a "bioneer" in researching the cause and effects of Morgellons Disease. At the turn of the 21st century he established a definite relationship between biologicals found in humans and aerosol spraying. Morgellons sufferers described painful sensations of insects crawling on, biting and stinging their skin.

According to studies by Albarelli and Martell, the sensation results in skin lesions that can appear like acne. They can appear anywhere in a patient's body and quite often contain fiber-like strands or fibrous material. When attempting to remove the fiber strands, the material will resist and act to withdraw or move away from the instrument being used for the extraction.

Detoxing heavy metal pollution in our bodies by eating only organic foods naturally strengthens our immune system and is essential to preventing many of the diseases listed above caused by decades of geoengineering and what has become known today as *"Hyper-Toxicity Syndrome."*

Weather Derivatives

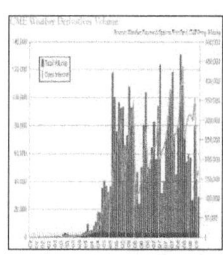

As aerosol spraying continues more people become sickened, more people depend on doctors and hospitals, and more money is made for Big Pharma who benefits from their incapacitation. Additionally, the powers-in-charge end up with a population unable to protest against geoengineering practices. It is a win-win for those who have wielded command and control over the population for so long. In fact, market future trading is being conducted in weather related derivatives today. Former Commerce Secretary William Daley testified to Congress in 1998 that "Weather is not just an environmental issue; it is a major economic factor. At least one trillion dollars of our economy is weather sensitive."

Trivia: The reason a bull and bear represent market direction is because of they way they attack. A bull will charge up with his horns and a bear will claw down, hence we have bull and bear markets.

Weather Derivatives Companies

CME Group
AccuWeather
Artemis
AWS Energy Services
Climate Prediction Center (CPC)
EarthSat
EQECAT
Energy.net
Evolution Markets
First Enercast Financial
Galileo Weather
GuaranteedWeather
ICAP
IPS MeteoStar
Spectron Group
Speedwell Weather Derivatives (SWD)
Thompson Reuters / Insurance Linked Securities Community
Tradition Financial Services (TFS)
Vyapar Capital Market Partners
Weather 2000
Weather Bug
Weather Insight
Weather Risk Management Associate (WRMA)
World Climate Service
Weather Service International (WSI)

Imagine how much profit a small group could make if they controlled the weather at will? Imagine the power one would have if they could wirelessly control mind and body remotely using geoengineering, HAARP radar, supercomputers, nano-technology, and you! This is not science fiction but decades old technology being advanced at the highest levels by the powers that must cease to be. You can read more about this, another primary use of geoengineering operations, in Book III titled, *"Touch-less Torture; Target Humanity."*

Below is a description from Elana Freeland's book, *"Chemtrails, HAARP, and the Full Spectrum Dominance of Planet Earth,"* describing the process used for the GE Hurricane Sandy event:

"Before landfall on October 29, 2012, nano-particulate aerosols were dumped from 40,000 feet into the hurricane core. Satellite Doppler indicates two layers of clouds, one of natural thunderheads, the other chemtrail-created....the hurricane was manipulated under Operation HAMP (Hurricane Aerosol Microphysics Program) of the Department of Homeland Security (DHS). HAMP operates in tandem with the DHS WISDOM Project (Weather In-Situ Deployment Optimization Method), Unmanned Aerospace Systems (UAS) projects (drones), and National Oceanic Atmospheric Association (NOAA)."

Sandy appears to have been an Atlantic Ocean experiment to test NASA's Hurricane and Severe Storm Sentinel (HS3). The purpose to address "the controversial role of the Saharan Air Layer (SAL) in tropical storm formations and intensification as well as the role of deep convection in the inner-core region of storms," needing, "sustained measurements over several years." The more data points they collect the more accurate their abilities to control weather.

Another benefit is the renewal and rebuilding of affected areas for corporate and private for-profit enterprises, damn the displaced refugees that are still trying to recover from their loss of money and property.

Below is a partial list of corporations that are contracted out to accommodate weather modification agendas.

For-Profit Weather Modification Private and Quasi-Government Corporations

Advanced Radar Corp
Aero Systems, Inc.
Aerotec Argentina
Colorado Water Conservation Board
Deepwater Chemicals
Direccion de Agricultura y Contingencias Climaticas (Argentina)
Droplet Measurement Technologies
Dynamic Aviation Group, Inc.
Electronic Systems Development C.C.
General Aviation Applications, 3D S.A.
Hydro-Tasmania
Ice Crystal Engineering, L.L.C.
Idaho Power Company
NASIC/DEKA–US Air Force Wright-Patterson AFB North American Weather Consultants, Inc.
North Dakota Atmospheric Resources Board
Omni International, L.L.C.
Radiometrics Corporation
R.H.S. Consulting, Ltd.
Sacramento Municipal Utility District
Santa Barbara County Water Agency
Snowy Hydro Limited
Southern California Edison County
National University of Technology, Mendoza, Argentina
Utah Division of Water Resources
Vaisala.com
Vaisala Oyj
Vaisala, Veriteq
Weather Modification, Inc.
Western Kansas Groundwater, District 1
Western Weather Consultants, L.L.C.
Wyoming Water Development

Texas Weather Modification Association (TWMA)

West Texas Weather Modification Association
Southwest Texas Rain Enhancement Association
South Texas Weather Modification Association
Panhandle GCD Precipitation Enhancement Program
Southern Ogallala Aquifer Rain Program (SOAR)
Seeding Operations and Atmospheric Research (SOAR), old website
The Edwards Aquifer
Colorado River Municipal Water District, Engineering, WXMOD
Trans Pecos Weather Modification Association

Military

USAF Reserve, Aerial Spray Unit, Youngstown A.F.B.
NASIC/DEKA, United States Air Force Wright-Patterson A.F.B.

Other

Aquiess and Drake International, Global Rain Project
Meteo Systems, Weathertec
Australian Rain Technologies, ATLANT
Ionogenics, ELAT
Evergreen Aviation: Supertanker
Kansas Water Office

Chapter 2 | Geoengineering 101—Who, What, Where, How, and Why

"We will know our disinformation program is complete when everything the American public believes is false."
—William Casey, CI Director (1981)

Programming the mind began before the days of Copernicus in the 16th century. It was used to create a disillusion of what people actually saw.

Pointing out obvious "persistent contrails," as many in the scientific community refer to geoengineering operations can be frustrating to people who want to alert their family, friends, and co-workers to what is hidden in plain sight in the skies above. Most do not want to deal with this critical reality even when the evidence is literally right in front of them. It is largely due to mass media's success in convincing us that the sky trails are just contrails and occur because of atmospheric temperature changes.

Factoid: "Conspiracy theory" was introduced into the minds of the programmed masses directly after the completely fraudulent Warren Commission reports came out to diffuse and obfuscate another in the long line of lone gunman assassinations of famous people.

Today's modern jet engines are nearly incapable of producing condensation trails except under extreme conditions, yet confusing language is used to dissuade and deflect attention away from geoengineering. This is known in psychology terms as cognitive dissonance.

Years of PSYOPS have convinced the public that jet exhaust causes lingering "contrails" anti-GE activists have had difficulty persuading the uninformed that what the military labels "chaff" is really deliberate, persistent chemical spraying. The ratio of air-to-exhaust is much too high to facilitate the formation of condensation because the majority of air expelled from the back of the engine is not combusted. It is passed through the "fan" and simply blown out the back without mixing with any fuel at all.

Turbine engines are the power plant for high-bypass turbofans. These engines are used in other applications besides powering jets. They are also used to power helicopters and many prop driven planes, yet we never see trails coming from these types of vehicles and the reason is simple. Turbine engines virtually never produce condensation trails.

When the unaware populace doesn't realize they're being sprayed like insects, the instant response is: *"Why would they spray their own children?"* This is an excellent question, which has no real answer other than perhaps:

- They are not humans doing this and they are terra-forming Earth to their own needs.
- They are royal blue bloods and are either immune to the toxic effects and or get chelation blood treatment therapy to remove the heavy metal toxins we have inside our bodies. We have living parasites as well as nano-particulate matter inside of us, as well-documented by the German scientist, Harald Kautz-Vella in his "Base" series of lectures on YouTube.
- The third possible reason the power-elite today would choose to kill all life on Earth so willingly and deliberately over decades is that they are sociopaths and psychopaths and care not one-cent about Life itself, human or otherwise.

Wall Street makes the most money when resources get the most scarce. There is no soul, no morality, and its "just business" to capitalize for maximum profits no matter what the future costs. A prime example is the $20 trillion of debt the US has accumulated in just the past decade where our children and likely their children will owe and owe and owe. It is also why our government refers to us as "human resources."

When I began a career on Wall Street in the mid-1980's, one of the first people to speak to me on my first working day as clerk on the New York Stock Exchange, was a senior partner at my new firm. He spoke

Chapter 2 | Geoengineering 101—Who, What, Where, How, and Why

directly and plainly, while sitting at the 7th floor bar above the NYSE, where brokers go to unwind. He said, "Son, if you are going to make it in this business you must know two things about Wall Street truths." He looked me straight in the eye and said, "You're either at the table or on the table" and "Here on the Street, you're either prey or you pray." That was over 30 years ago and greed has now been declared "good" no matter what the cost.

Wall Street is owned by the "banksters," who are owned by the families of Court Hofjuden Jews who report to the Bank of International Settlements (BIS) located in Basil, Switzerland. The BIS is a clearing house for, and controls all central banks, including the Federal Reserve. When you read about the British exiting from the EU, it's all theater—follow the money, no matter what the country, the buck stops at the House of Rothschild—the Jews who clear for the Vatican. Switzerland, is a host to the Rothschild's banking empire and secret bank accounts. Switzerland has always stood alone, always neutral and never invaded through two world wars. Every citizen carries and knows how to use a rifle. Swiss guards provide Vatican security and the Rothschilds are the Vatican's personal bankers. Switzerland has a Red Cross for a flag that is also embedded in the flag of Great Britain.

Why would the US declare independence from Britain in 1776 in a fight to the death and then choose the exact same colors for their flag as the country they just defeated? When the Soviet Union collapsed, the new flag was also red, white, and blue, as well as the flags of Chile, Puerto Rico, France, Costa Rica, Australia, the Netherlands, and the East Indian Tea Company.

The "bankster's" rulers are the Society of Jesus, also known as the Jesuits who are headed by the current Black Pope, Father Adolfo Nicolas. (He is six levels above the current White Pope Francis, the first Jesuit pope in the Vatican in 600 years). Pope Francis was Archbishop of Argentina during the 1980s "Dirty Wars" where tens of thousands of Argentinians simply "disappeared," though he has claimed ignorance regarding the purge of human life.

One cursory look at the "Kitchen Cabinet" staff around our current President, Barrack Obama, and you begin to understand who is REALLY in control and has been for a very long time.

Some of the key officials during President Obama's campaign and in his cabinet include his top speech writer, Jon Favreau, who trained at the College of the Holy Cross in Worcester, Massachusetts, the oldest Jesuit College in New England. Dan Pfeiffer, who was Obama's Deputy Communications Director during the campaign and continues in the same position at the White House, graduated from Georgetown University.

His senior Military and Foreign Policy Advisor was Major General Jonathan Scott Gration, a fighter pilot whose masters' degree is from Georgetown University, the oldest Jesuit institution in America. Most of the top brass of the Joint Chiefs of Staff graduated from the Edmund A. Walsh School of Foreign Service.

Bill and Hilary Clinton graduated from Georgetown University and Donald Trump went to Fordham University, a Jesuit school.

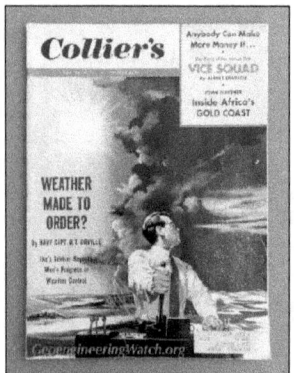

Collier's Magazine, *"Weather Made to Order"*, May 1954

The Governor of California, Jerry Brown, is a well known Jesuit as is his successor, former Mayor of San Francisco, Gavin Newsom, who is a graduate of Santa Clara University, a Jesuit school. Janet Napolitano, Head Regent for the University of California schools and former head of the Department of Homeland Security (DHS) graduated from Santa Clara University as well; just another coincidence, I guess.

"The Office of the Gene Technology Regulator (OGTR) is close to approving a license application from PaxVax Australia (PaxVax) for intentional release of a GMO vaccine consisting of live bacteria into the air over Queensland, South Australia, Western Australia and Victoria." According to Australia's regulators, this human aerosol Geo-gene engineering qualifies as a limited and controlled release under section 50A of the Gene Technology Act 2000:

"(2014) Drug company, PaxVax, has sought approval to conduct the clinical trial of a genetically modified live bacterial vaccine against cholera. Once underway the trial is expected to be completed within one year, with trial sites selected from local government areas (LGAs) in Queensland, South Australia, Victoria, and Western Australia. PaxVax has proposed a number of control measures they say will restrict the spread and persistence of the GM vaccine and its introduced genetic material yet the regulators readily admit they have little knowledge as to how the spraying effects plants, trees and wildlife."

ICAS Report No. 10a
November 1966

A Recommended National Program In Weather Modification

A Report to the

Interdepartmental Committee for
Atmospheric Sciences

by

Homer E. Newell

Associate Administrator for Space Science and Application
National Aeronautics & Space Administration
Washington, D.C.

*Interdepartmental Committee
for
Atmospheric Sciences*

FEDERAL COUNCIL FOR SCIENCE AND TECHNOLOGY

Executive Office of the President

Chapter 3 | Aerosol Spraying of Vaccines Ramps up World-Wide

Weather modification will become a part of domestic and international security and could be done unilaterally... It could have offensive and defensive applications and even be used for deterrence purposes. The ability to generate precipitation, fog, and storms on earth or to modify space weather, ... and the production of artificial weather all are a part of an integrated set of technologies which can provide substantial increase in US, or degraded capability in an adversary, to achieve global awareness, reach, and power.

—*US Air Force, [emphasis added].
Air University of the US Air Force, AF 2025 Final Report*

In the 1920s in Puerto Rico, John D. Rockefeller funded sterilization of hospitalized women without their knowledge or consent. In the 1980s Rockefeller funded GMO campaigns in Argentina before being introduced world-wide. It now appears that Australia is prepared to launch GMO vaccine aerosol spraying on its people and it's legal and documented; damn the knowledge or consent of unknown implications to the health and well being of the populace.

In August of 1996, the US Air Force created a document entitled, "*Weather as a Force Multiplier: Owning the Weather in 2025.*" They even included the USAF cadet training classes in the project. The military has unlimited funds for projects that fit into their overall Command and Control strategy for what the military calls "Full Spectrum Dominance." Keep in mind all of this is funded by ignorant American tax payers who could never comprehend that the US military would plan such attacks on its citizens— think again.

In 2006, Michael Greenwood wrote an article entitled, *"Aerial Spraying Effectively Reduces Incidence of West Nile Virus (WNV) in Humans."* The article stated that the incidence of human West Nile virus cases can be significantly reduced through large scale aerial spraying that targets adult mosquitoes, according to research by the Yale School of Public Health and the California Department of Public Health.

Under the mandate for aerial spraying for specific vectors that pose a threat to human health, aerial vaccines known as DNA Vaccine Enhancements and Recombinant Vaccine against WNV, are being tested and are being used to "protect" the people from vector infection exposures. In other words, these sprays can, are designed, to alter ones own DNA. *Are you fully creeped out yet and fully understand all that this implies for command and control of humanity?*

Factoid: Congressman Dennis Kucinich attempted to halt geoengineering in 2001 through Bill HR2977. Two years later the "Space Preservation Act" was passed but excluded any mention of geoengineering practices.

Can you begin to understand the near and long term implications using the human population as unsuspecting lab rats, where aerosol spraying will make us ill, unruly, or laconic at will?

Government and corporate funded papers from Harvard, Yale, Oxford, and Berkeley, etc., what many consider to be prestigious and reputable institutions are used to validate the climate-weather engineers as well as invalidate claims that the government would consider using aerosol spraying on the public. Politicians are willfully powerless to address the ongoing ecocide and ecophagy occurring almost daily on everyone around the world.

These quasi-public-scientific documents have lead directly to legitimize aerosol spraying. In the fall of 2007, heavy spraying was conducted around Northern California, allegedly to eradicate the light brown moth from ruining farmer crops in the Carmel Valley of Monterey County. The Bay Area has had under-the-radar mosquito abatement spraying programs for decades since many live adjacent to waterways but reality caught up with the NIMBY crowd when the elite Carmel Valley area was dosed in 2007.

Over several months that year, many developed respiratory illnesses as a direct result of the intensified spraying. Calling your Congress person did nothing. Air Quality boards were silent. Law suits were filed and nothing stopped until the program was completed or went under-the-radar with less visible dispersants.

Vaccines, Geoengineering, and You

"The world has 6.8 billion people...that is headed to 9 billion. Now if we do a really good job on new vaccines, health care, reproductive health services. We could lower that by 10 to 15 percent." —William Gates, *Eugenicist, Population Control Expert, World Vaccine Administrator, Common Core Education Director, Member of the Trilateral Commission, Bilderberg Group, Committee of 300, Pilgrim Society and Philanthropist.*

The wealthy elite, including Sir Richard Branson of Virgin Airlines, Skype co-founder, Niklas Zennstrom, and the Bill and Melinda Gates Foundation have covertly funded geoengineering proponents like David Keith of Harvard University and Ken Caldeira at Stanford University. According to financial statements, Mr. Keith receives undisclosed sums from Mr. Gates each year and is president and majority owner of a geoengineering company, Carbon Engineering, in which Mr. Gates and friends are also partners. No conflict of interests here!

It is no coincidence that Bill Gates has lead the charge for artificially intelligent (AI) vaccine-carrying mosquitoes to treat Malaria, that he first demonstrated on an unsuspecting audience at a TED talk he gave in 2009. Now, just over a half-decade later, we learn that Gates and company had released AI mosquitoes into the exact same area where the alleged Zika virus is coming from in Brazil.

It was reported in June, 2016 that helicopters were spraying areas over the five boroughs of New York City to arrest the Zika virus and airplanes were injecting vaccine into flights to and from Brazil without anyone's awareness. By July, NBC news reported the following quote came from the Director of the Center for Disease Control (CDC), Dr. Tom Frieden:

"If any part of the continental US had the kind of spread of Zika that Puerto Rico has now, they would have sprayed months ago. This is more a question of neglect than anything else." He went on to say that the product "Naled" was used last year on 6 million acres in Florida, including Miami as well as New York without any increase in asthma cases there. Incredibly the public has never been informed of the illegal activities and civil rights infringements on being treated like insects in government lab experiments.

Biometric IDs

Five US states now have requirements that all citizens in those states get a biometric REAL ID or they will not allowed to fly commercially in the States. It is possible that *every* US citizen will be required to have a biometric REAL ID and you must show evidence that you were vaccinated if you wish to fly anytime in the future. If this goes unchecked then access to hospitals, banks, public parks, sports venues, etc., won't be allowed until everyone is inoculated. This is in addition to being aerosol-sprayed from above.

Command and Control is their creed.
Do What thou Whilst is their motto.
Christians and heretics alike must
be exterminated is their mission

Bill Gates is the poster child for evil in the public domain. It is easy to see the vast amount of evidence of death and destruction, which so often presented as altruistic and benign. When 47,500 people in India were paralyzed by Gates, who introduced a polio vaccine in 2009, he shrugged it off but did not deny the allegations instead focusing on the benefits the vaccine protocol had produced. The same language was used in Pakistan

Chapter 3 | Aerosol Spraying of Vaccines Ramps up World-Wide

where Gates and company were directly blamed for 10,000 related deaths in 2013. This is another "too big to jail" club of murderers wishing the population be reduced by some 92% as declared in the "Georgia Guidestones."

As hard as it is for most of us to comprehend or believe, the wealthy elite care nothing about you, your children, or other Life on Earth. We are human resources to be exploited. In Roman Latin legalese we are referred to as "a property or thing" and as "chattel" much like the word "cattle."

Gates's wealthy donors protect the interests only for those who make up the behemoth Bill and Linda Gates Foundation. More than $70 billion, tens, perhaps hundreds of millions of dollars go to promoting *geoengineering*, GMO vaccine spraying, and Artificial Intelligent drones with payloads of their choosing.

Gates has also declared publicly that there are more than enough people in the world. His father has been heavily involved in the Rockefeller founded and funded Planned Parenthood. He has boldly declared at other TED talks that if there were better methods, some including vaccinations, they could get the population down 10%–15%. Apparently Gates feels there are enough people in the World. He has put himself in charge of solving the problem that he and his Club of Rome cronies helped create by promoting endless consumption and causing massive consumer debt partly by using Asian slave labor to produce goods because, as Dick Cheney said infamously, "the American lifestyle is non-negotiable." President Bush famously declared directly after the self-directed 9/11 attacks that we should all feel better and "go shopping." Does anyone bother to ask what will happen when continuous consumption causes Earth to become so toxic she must sweat us out to eliminate (us) parasites?

> Those who worry about being chipped do not realize we are all already chipped, monitored and controllable already. There is no need for external chipping devices when vaccines and now, aerosol application vaccines, easily hold the necessary command and control devices.

Geoengineering Not Chemtrails:
Investigation Into Humanities 6th Great Extinction

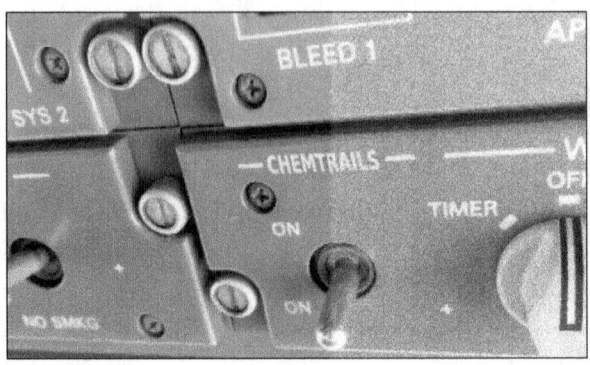

Chapter 4 | Get Your College Degree in Geoengineering

"But there are at least 26 reasons why geoengineering may be a bad idea. These include disruption of the Asian and African summer monsoons, reducing precipitation to the food supply for billions of people; ozone depletion; no more blue skies; reduction of solar power; and rapid global warming if it stops." —Alan Robock, Professor, Department of Environmental Sciences, Rutgers University, 2013
(The Oxford Geoengineering Programme was founded in 2010 as an initiative of the Oxford Martin School at the University of Oxford).

Geoengineering– The deliberate large-scale intervention in the Earth's natural systems to counteract climate change is a contentious subject and rightly so.

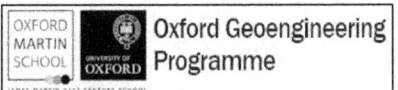

The Oxford Geoengineering Programme seeks to engage with society about the issues associated with geoengineering and conduct research into some of the proposed techniques. The program does not advocate implementing geoengineering but it does advocate conducting research into the social, ethical, and technical aspects of geoengineering. This research must be conducted in a transparent and socially informed manner.

The University of Oxford is involved in three major projects on geoengineering funded by the UK Research Councils. They are: the Integrated Assessment of Geoengineering Proposals (IAGP) in partnership with The University of Leeds, Cardiff University, Lancaster University, University of Bristol, University of East Anglia, the Tyndall Centre, and the UK Metropolitan Office; Stratospheric Particle Injection for Climate Engineering (SPICE) in partnership with the University of Bristol and Cambridge University and Climate Geoengineering Governance (CGG). A recently announced Oxford-led project in partnership with the University of Sussex and University College London which will examine the governance and ethics of geoengineering [*emphasis added*].

Harvard University, Center for the Environment

The Working Group: Solar Geoengineering is the concept of deliberately cooling the Earth by reflecting a small amount of inbound sunlight back into space. This seminar series held jointly by the Harvard University Center for the Environment (HUCE) and MIT's joint program on Science and Policy of Global Change will explore the science, technology, governance, and ethics of solar geoengineering. In bringing together international experts, participants will learn some of the greatest challenges and hear opinions on how this technology could and should be managed.

University of Texas, Austin

The Center for Integrated Earth System Science (CIESS) is a cooperative effort between the Jackson School of Geosciences and the Cockrell School of Engineering. The Center for Integrated Earth System Science (CIESS) seeks a deeper understanding of the physical chemical, biological, and human interactions that determine the past, present, and future states of Earth.

Specifically, the goal of CIESS is to answer a wide variety of earth science questions including:

- How can we use in situ measurements, global satellite observations, proxy data, and computational analysis to describe and understand Earth's dynamic system?
- What has been the impact of human activity on Earth?
- What is the future of our environment under climate change, land use change, and water use change?
- How can we reduce modeling uncertainties and make reliable predictions of extreme events at regional scales? How can we make rational decisions under uncertainties in order to mitigate, prevent, and plan for or adapt to negative potential impacts of global change?

Columbia University

A complementary degree (Master of Arts in Climate and Society) is available through Columbia University for students who are more directly interested in social or planning aspects of climate impacts, and are not quantitatively oriented.

University of California at Berkeley

Geoengineering is an interdisciplinary program that offers excellent opportunities for students with background in Engineering and Earth Sciences who are interested in all aspects of soil and rock mass characterization, development of advanced simulation techniques, performance of earth structures and underground space, and identification and mitigation of natural hazards.

University of Michigan

"For cloud seeding—we are already doing that. We have satellite images of ship-tracks available that tell us that in fact the emissions from the ships brighten the clouds in the little lines you see (chemtrails) and change the radius of the particles in the clouds. So we're already doing that." "....we also know that the aerosols that we are currently emitting are already protecting us from global warming to a certain extent...We have already seeded marine clouds," Professor Joyce Penner, University of Michigan.

Do these look like naturally occurring weather patterns?

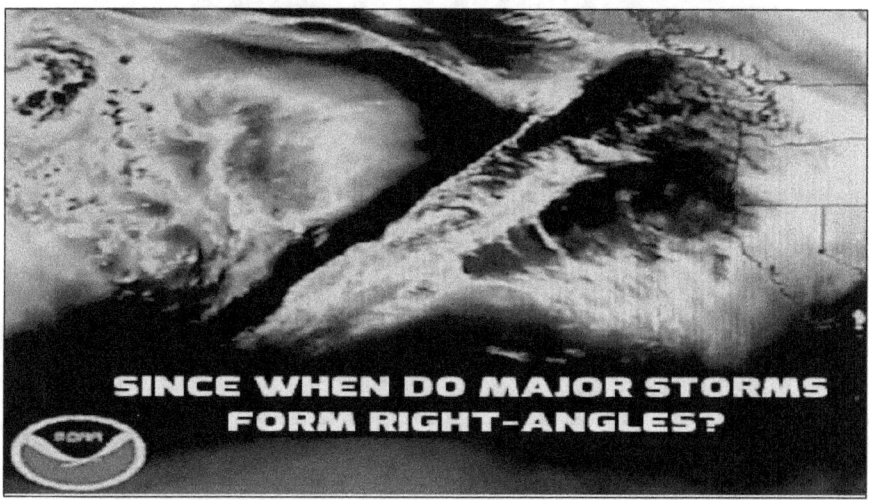

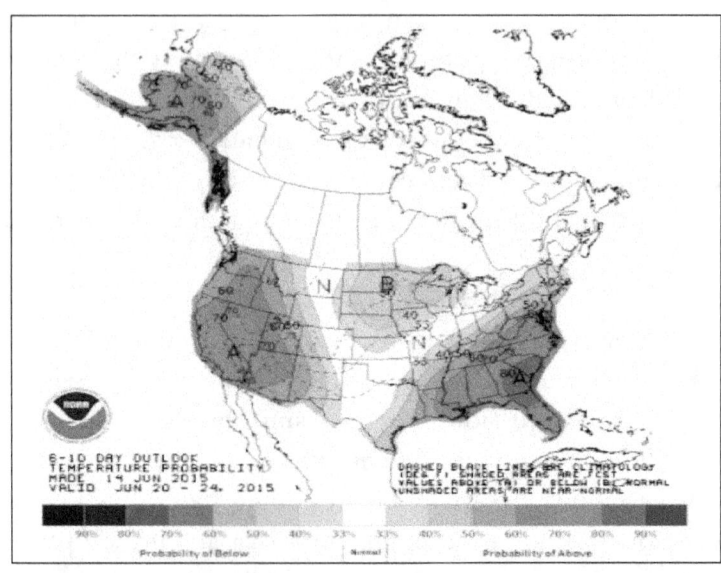

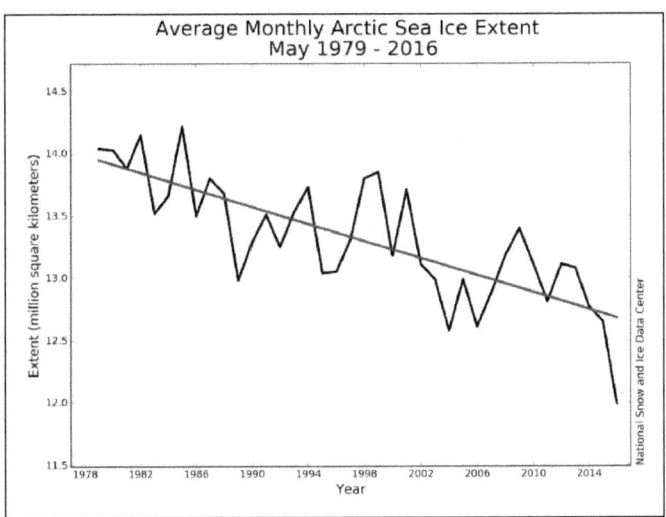

The arctic is on fire and thats not good for life on earth.

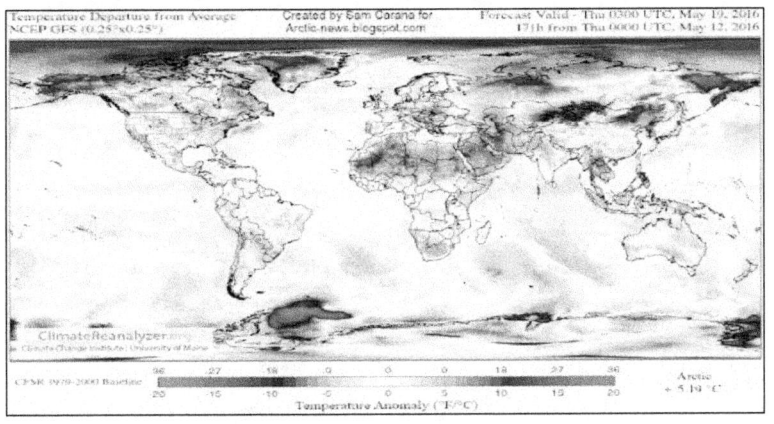

Chapter 5 | **The Great Arctic Melt Off**

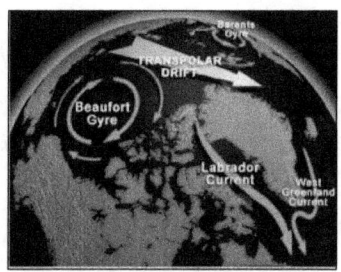

Arctic ice free by 2016! "38 Years ahead of our estimates"
—US Navy 2013
"Given that the Arctic is warming faster than the rest of the globe, understanding the processes and feedbacks of this polar amplification is a top priority."   *—US Department of Energy*

> *"The study estimates that in 2008, the US poured 49 million tons of methane into the air. That means US methane emissions trapped about as much heat as all the carbon dioxide pollution coming from cars, trucks, and planes in the country in six months"*

The last time the Arctic was ice free was 3.6 million years ago. Today, ice core models chronicling Earth's past do not even come close to measuring the Arctic liquidation occurring in the Great White Northern regions today. Alaska is averaging as of Spring 2016 temperatures consistently at 10–20°F above average what used to be considered normal for over the past century of recorded history. Snow actually had to be trucked in so the famous Alaskan Iditarod could be completed in temperatures of 50 degrees.

The Arctic Gyre steers the weather around the world like a swizzle stick above the Earth, which is helping to intensify the warming of oceans worldwide. One has to have their head completely in the sand to not be aware of the world-wide devastation of ocean life due to warming seas. Fish are found radiated and filled with mercury poisoning. Little reporting is done regarding the ongoing events from the Fukishima disaster and the deliberate ionization of our oceans by governments and private enterprises.

"Furthermore Professor Jennifer Francis has shown that the present CO_2 content of the atmosphere has a delayed temperature anomaly more than 12°C (53.6°F), which is higher than the Major Permian Extinction Event (Wignall, 2009) so we will be facing total extinction unless we sharply reduce our carbon dioxide emissions by a large amount (more than 90%) and the existing methane content of the atmosphere by more than 60%." —Professor J. Francis, Research Director of Marine and Coastal Sciences at Rutgers University.

The Most Important Chart You Will Ever See in Your Lifetime

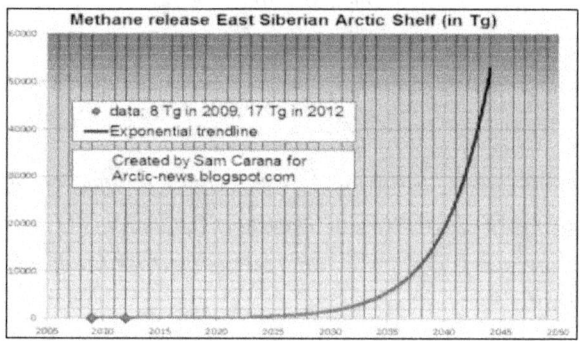

In 2014 the US Navy released an extensive report predicting that the Arctic could very well be ice-free by 2016, 84 years ahead of all other climate models. Previously, an exhaustive EPA study was conducted in 2008, which found methane gas releases were an astounding 50% higher than previous estimates due to fracking.

"The study estimates that in 2008, the US poured 49 million tons of methane into the air. That means US methane emissions trapped about as much heat as all the carbon dioxide pollution coming from cars, trucks, and planes in the country in six months."

That's more than the 32 million tons estimated by the US Environmental Protection Administration or the nearly 29 million tons reckoned by the European Commission. This study, however, was based on nearly 13,000 measurements from airplane flights and tall towers, the most used in any

Chapter 5 | The Great Arctic Melt Off

such research. Methane gas releases have gotten so bad that high-altitude planes clearly show methane release pools due to fracking practices.

In the fall of 2015, a massive methane release occurred at Porter Ranch in the Central Valley of Southern California. An astounding 3,346 tons of methane release into our atmosphere was reported per day! The situation got so bad the FAA put restrictions on flights in and around the LA Basin so planes engines would not penetrate the methane gas and explode in mid air.

Thousands of people living nearby had to evacuate and the entire LA Basin was saturated with deadly methane gases. Huge pools of methane gas have settled over the entire west coast of the US that will further accelerate the warming of an already very dry West.

"Something is very much off in the inventories," said study co-author Anna Michalak, an Earth scientist at the Carnegie Institution for Science in Stanford, Calif. "The total US impact on the world's energy budget is different than we thought and it's worse." What those failing to acknowledge is the increased methane gas releases are coming down from the Arctic.

Methane gas contains 20 times more CO_2 than other greenhouse gases and scientists have been extremely concerned about the acceleration of releases. Since 2007, greenhouse gases coming from the Arctic alone make up 10% of the total releases in North America today, yet that is nothing compared to what is in store for us if melting in the Arctic continues along its current trajectory.

In 2013, another large release of methane occurred in the Arctic regions and current reports are being suppressed that show the entire snow cap has melted and full releases of methane are now occurring. Methane gas takes 1½–2 years to fully deploy before the full greenhouse umbrella takes effect.

We are now in an irreversible trend towards what scientific scenario calls the "Venus syndrome." Venus syndrome is a scenario in which climate and atmospheric feedback loops are triggered that can't be switched off. Under this scenario, as greenhouse gases build up and cause world-wide warming, more greenhouse gases are released, which causes still more warming. This trajectory does not end in a balmy tropical resort type Earth, but rather a planet that is closer to hell. Like Venus, Earth would become a pressure cooking inferno with virtually no life. Geoengineering is only accelerating the certainty of this occurring over the next decade if not sooner as record warm temperatures are recorded across the Earth each and every year.

If "cooling" the planet was truly the primary goal of the geoengineering programs, their method is insane at best. The "pharmaceutical" approach, applying a "cure" is exponentially worse than the ailment it was meant to treat. It is increasingly evident that more straight forward goals are being carried out with the weather modification, weather warfare programs. Ultimate power and control are inevitably at the root of such programs. All available data indicates that if geoengineering programs continue, this "power" over the weather, and Earth's inhabitants will soon be at the cost of all life. Geoengineering can and does create large scale cooling events by blotting out the sun on an immense scale and with artificial ice nucleation of clouds and storms but it comes at the cost of much worsened warming overall. In addition to the consequences listed above, simply stated, geoengineering particulates and the toxic atmospheric haze they create traps more heat than it deflects. Our already dire situation worsens by the day.

The More Ice Melts, the Exponentially Faster Methane is Released into the Air.

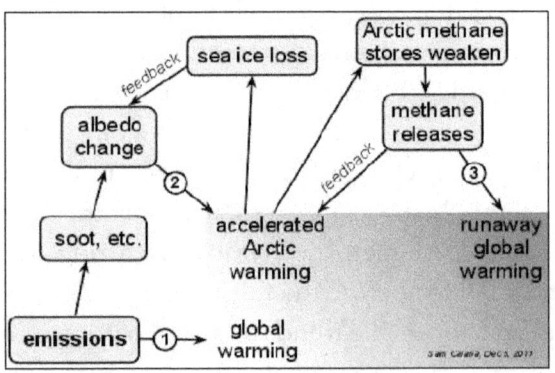

"We as a species have never experienced 400 parts per million of carbon dioxide in the atmosphere," Guy McPherson, *professor emeritus of evolutionary biology, natural resources, and ecology at the University of Arizona and a climate change expert of 25 years, told me. "We've never been on a planet with no Arctic ice, and we will hit the average of 400 ppm…within the next couple of years. At that time, we'll also see the loss of Arctic ice in the summers…This planet has not experienced an ice-free Arctic for at least the last three million years."*   —Guy McPherson, AMEG

Chapter 5 | The Great Arctic Melt Off

Oceans are being heavily damaged by iron fertilization in more desperate attempts to mitigate accelerated warming, sea life die-off, and massive pools of methane releasing from the ocean floor.

What is critical to understand is that the Arctic weather conveyor belt acts as a governing regulator for the entire Earth weather system and it is changing rapidly due to melting of ice sheets, especially along the Eastern Siberian Arctic Shelf. As more ice melts, less sun is reflected back into space, which means more accelerated melting will continue to occur. This is called the Albedo Effect.

Over the past few years world powers have been actively engineering an all out world-wide aerosol campaign experiment in a desperately, ill-fated effort to 're-winterize' the Arctic with geoengineered, chemically nucleated-treated snow.

There is large-scale aerial spraying occurring in the Arctic region with heavy bombardments of chemicals utilizing high-tech practices including massive doses of aluminum, barium, and strontium. Additionally, while using HAARP towers to steer winter weather destined for the US West Coast to help with the Re-Arctification efforts, California's watershed is being sacrificed, where over 40% of agriculture is grown to feed the US.

Only recently did the best scientists and the US Navy begin to understand that once the Arctic tundra gets exposed it will cause a massive release of million years of deadly methane gases from once-frozen, buried pools.

At this time they decided to shift their attention to nucleating the Arctic with aerosol spraying (steering weather using HAARP coupled with geoengineering the skies above) in an increasingly faulty and desperate attempt to refreeze the Arctic to keep methane sealed under ice and snow. This is a major factor for causing the intense, chaotic and geoengineered weather over the past several years. We are currently seeing huge, daily swings in temperatures. The 2014 Polar Vortex that swept over the Midwest and Eastern US is causing record snow and low temperatures as well as disputing the fact that Earth is burning up everywhere else. It directly puts the entire Earth's weather systems in absolute chaos.

Government, military, and private enterprise have been laying down thick layers of iron for years, trying to re-ionize the fecundity of the oceans without success, while helping to push the mass migration northward and cause die off of fish along the Pacific Ocean coastline.

Ecosystem Die-off Watch

All along the west coast of Northern California, we are experiencing a complete loss of salmon and trout in our streams over the past decade because of warming rivers, clear cut logging, and general apathy for the complete destruction of their life giving habitat.

Marine biologists up and down the coasts report dead sea lions, whales, and other large creatures foundering and dying in unseen numbers. Feeder krill counts that provide critical food supplies for the ocean life food chain are at record lows as ocean water temperatures continue to hit record highs. Crab season all up and down the west coast of has been canceled again this year due to toxicity. It has gotten so bad that in 2015, the Chinese announced they would not be purchasing any shell fish from the US any longer.

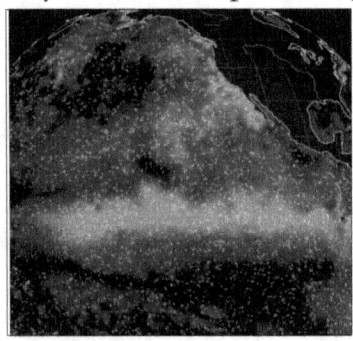
Massive methane gases hovering over entire Pacific West Coast, 2015

In April of 2016, marine biologists reported unprecedented large-scale die-offs of the sea forest kelp. This from the Santa Rosa Press Democrat, *"Collapse of Kelp Forest Imperils North Coast Ocean Systems."* (April 2016). "Large tracts of kelp forest that once blanketed the sea off the North Coast have vanished over the past two years, a startling transformation that scientists say stems from rapid ecological change and has potentially far-reaching impacts including on several valuable fisheries. The unprecedented collapse has been observed along hundreds of miles of coastline from San Francisco to Oregon. The region's once-lush stands of bull kelp, a large brown algae that provides food and habitat for a host of wildlife species have been devoured by small, voracious purple urchins. In the most-affected areas denuded kelp stalks are almost all that remains of plant life."

"There's almost no rock-fish species that is not of economic importance," said Milton Love, a rock-fish expert and associate research biologist at University of California, Santa Barbara's Marine Science Institute. The young of many rock-fish species also are a major food source for seabirds, salmon, and some pinnipeds, like sea lions. A major collapse of those populations could snap the food chain. *"It's very concerning,"* Catton said. *"There's a lot of uncertainty moving forward."*

Chapter 5 | The Great Arctic Melt Off

"The kelp collapse may only add to the woes. Purple urchins are silver-dollar-sized species rarely caught for commercial harvest in California. They normally co-exist in kelp forests alongside other marine life, including red urchins....and red abalone. But those two species also feed on kelp and both are showing signs of starvation," Catton said.

"Rock-fish, another key fishery that includes dozens of species sought by commercial and sport anglers, also likely will take a hit from the kelp die-off," said Mark Carr, a professor of ecology and evolutionary biology at University of California at Santa Cruz. "The young of several near-shore species take shelter in the kelp during their first months of life."

The excerpt below is from a report by Mr. Michael T. Snyder at www.theeconomicblog.com website:

"Why are millions upon millions of dead sea creatures suddenly washing up on beaches all over the world? It is certainly not unusual for fish and other inhabitants of our oceans to  die. This happens all the time but over the past month we have seen a series of extremely alarming mass death incidents all over the planet As you will see below, many of these mass death incidents have involved more than 30 tons of fish. In places such as Chile and Vietnam, it has already gotten to the level where it has become a major national crisis. People see their coastlines absolutely buried in dead sea creatures, and they are starting to freak out.

For example, just check out what is going on in Chile right now. The following comes from a Smithsonian Magazine article entitled *"Why are Chilean Beaches Covered with Dead Animals?"*

"Compared to other countries, Chile is almost all coast and that geographical fluke means that the country is known for its beautiful beaches. But that reputation may be on the wane thanks to a new sight on Chilean shores: dead animals. Lots of them. Heaps of them, in fact," as Giovanna Fleitas reports for the Agence France-Presse, "the South American country's beaches are covered with piles of dead sea creatures-scientists are trying to figure out why."

"Tales of dead animals washing up on shore are relatively common; after all, the ocean has a weird way of depositing its dead on shore. But Chile's problem is getting slightly out of hand." Fleitas writes, "recent months have not been kind to the Chilean coast, which has played host to washed-up carcasses of over 300 whales, 8,000 tons of sardines, and nearly 12 percent of the country's annual salmon catch, to name a few." Authorities in Chile are scrambling to come up with a reason why this is happening but nobody seems to be sure what is causing the tsunami of death. In Vietnam things are even worse. At this point so many dead fish and clams have been washing up along the coast that soldiers have been deployed to bury them.... Over and down-under at sea level in Australia not only are they—year after year—shattering record highs but native eucalyptus trees are dying in droves, corals reefs are dying in faster and faster time event horizons."

This reporting from David Sigston, AAP, May 10, 2016:

"Scientists are worried about an 'unprecedented' die-off of mangroves in northern Australia and the link with large-scale coral bleaching of the Great Barrier Reef. The widespread damage to mangroves around the Gulf of Carpentaria has been highlighted at an international wetland conference held this week in Darwin. While a detailed scientific survey has yet to be undertaken, photographs revealed hundreds of hectares of mangroves dying in two locations along both the west and east coastlines of the gulf." Professor Norman Duke, spokesman for Australian Mangrove and Saltmarsh Network said the scale and magnitude of the loss appears, "unprecedented and deeply concerning."

Professor Duke noted the damage was particularly alarming given this year's severe coral bleaching of the Great Barrier Reef as it appeared to correlate with extreme warming events in the region.

Until more extensive research is done, the James Cook University professor isn't sure if the mangroves are beyond saving but is warning more needs to be done. "Shoreline stability and fisheries values amongst other benefits of mangrove vegetation are under threat," he said.

The future ain't what it used to be. —Yogi Berra

Chapter 5 | The Great Arctic Melt Off

"A mysterious pathogen is wiping out starfish along the Pacific coast... [and it] isn't the only weird thing to happen of late along the California coast. Marine scientists have been trying to find out why previously unknown blooms of toxic algae are suddenly proliferating along the coast. The mysterious blooms including deadly red tides have been bigger, occurred more frequently and killed more wildlife than in the past. Last year at about this time legions of big predatory Humboldt squid gathered along the Northern California coast and stranded themselves on Santa Cruz beaches far north of their normal habitat," (*San Francisco Chronicle, 2013*)

In addition to our modern day version of "Oceans Gone Wild," there are world-wide reports of mass algae blooms occurring everywhere. Another interesting, coincidence theory not to be a conspiracy, connects President Obama's science adviser, John Holdren to announce in 2009, that the administration was advocating for the "fertilizing" of our oceans. Just a few months later US scientists announced in December 2013, that they had succeeded in producing crude oil from algae in under one hour. Big Oil is already harvesting the man-made toxic algae for fuel. Profit they will and must, accelerating our extinction event even faster!

In Northern California the Russian River was declared off limits due to toxicity of algae blooms in 2015. In the summer of 2016, Florida is also reporting never before seen algae blooms:

> In November of 2014, Mendocino County, Northern California became the first county in the United States to give legal Rights of Nature and passed new laws declaring county jurisdiction over state and federal laws. This enforceable law gives Mendocino County legal standing to protect Nature from egregious corporate malfeasance in the courts. Everywhere else Nature is only deemed a commodity to be plundered and sold. Additionally, Ecuador and Bolivia have also put similar Rights of Nature laws into their Constitutions. As we lose more and more of Nature's abundance and all-Life giving forces, it is imperative that we protect her legally from further destruction, since it is she alone, Mother Gaia, that gives Life to us all.

"Animals are in distress, some are dying, the smell is horrible due to the toxic algae bloom of green slime," Jordan Schwartz, owner of Ohana Surf Shop, reported from Stuart Beach off Florida's east coast. "This town is driven by tourism but the tourism is empty. You have to wear a mask in the marina and the river. It's heartbreaking and there is no end in sight." Florida Governor, Rick Brown, has declared a state of emergency in Martin, St. Lucie, Lee, and Palm Beach counties because of the toxic algae bloom that originated in Lake Okeechobee and spread to the beaches, CNN reports.

> *"Only when the last tree has died and the last river been poisoned and the last fish caught will we realize we cannot eat money."*
>
> —*Cree Indian Proverb*

Subliminal Messaging: Programming acceptance of GE into the subconciosness mind through advertisements

Chapter 6 | The Trees are Dying—We're Next

The Russian Hydro-meteorological Center recently reported that since May 2015, every single month has been the warmest in Russia's history. By way of example, in March, the temperature deviation on islands in the Barents Sea was a staggering 12°C (54°F)!

As of Spring 2016, Alaska's heat is breaking records by the dozens. Recent statements from the National Weather Service reported that the towns of McGrath and Delta Junction in the interior of the state hit a high of 78°F and a low of 49°F respectively, beating the previous records set in 2005 and 1988 for each. Fairbanks set a new high temperature record of 82°F which shattered a century-old record of 80°F set in 1915.

Anchorage, Alaska set a record of 72°F, a stunning seven degrees above the previous high that was set in 2014, while Juneau and Bethel set new heat records. Even Barrow in the far north, saw 42°F recently breaking the previous heat record by four degrees. Given that Anchorage has already seen the second-largest number of record high temperatures for any year and there is still most of the year to come, 2016 will certainly break the previous record of high temperatures seen, which was set in 2003.

In Africa, the heat continues to be unrelenting and that trend is expected to not only continue but increase according to a study recently published in the journal Environmental Research Letters. According to the study, by 2100 heat waves on that continent will be hotter, last longer, and occur with much greater frequency.

Southwest Asia and India recently saw historic heat waves that have caused more than 150 deaths. Cambodia and Laos each set record highs for any day of the year during April. Cambodia saw 108.7°F on April 15, and on April 26, Thailand set a record for national energy consumption (air conditioning) according to the Associated Press.

India went on to break its heat record in May when the city of Rajasthan saw 51°C (123.8°F) as the heat wave besetting northern India, which persists as temperatures have exceeded 40°C for several weeks in a row.

Years 2014 and 2015 were the warmest ever recorded and six months into 2016 we are set to beat those records,\; climate change "deny-ists" be damned.

One of the research team's authors said that "unusual" heat events will become much more regular, "meaning it can occur every year and not just once in 38 years—in climate change scenarios."

In 2015 a study from Stanford University reported in the journal of the *"Proceedings of the National Academy of Sciences,"* scientists are predicting we have "entered into a new era where nearly every year we have warming temperatures we will have low precipitation." The report goes on to say that "essentially all years are likely to be warm—or extremely warm—in California by the middle of the 21st century with less precipitation.

California is in the midst of the greatest drought in its history. For the third year in a row, Central Valley farmers have received little to no water allotments to grow food while neighborhood lawns are green, swimming pools are full, golf courses are green, and green money continues pouring into the coffers of government water agencies.

Californians aren't the only ones feeling the effects of our collective "new normal," hotter weather and precious little rain. In Colorado they have been consistently shattering record warm temperatures by some 5–15 degrees. Eastern Colorado had record high temperatures of 82°F in early February of 2015 and counties as far away as Los Angeles are being told the Colorado River will soon be unable to provide them with fresh drinking water.

The giant Redwood trees on the west coast of California is one of recent accelerated die-off as well. These great Sequoias have lived for hundreds and hundreds of years, yet less than 3% of these ancient forests remain today largely due to advanced destruction technology and capitalist greed in this era of Industrial *"Devilution."* Over the past few years, California has seen an unprecedented (*there's that word again*) extreme 5 year-long drought, where little fog now comes to cool and feed the giant Redwoods and moisten the loamy soil they thrive in. Only two ancient giant forests remain in the world today in South America and Africa.

Not only are we seeing the die-off of these ancient legends like never before but we are also getting reports around the world of enormous die-off of all species of trees from Australia to Bulgaria to Southern California.

Literally everywhere.

Chapter 6 | The Trees are Dying—We're Next

Rain forests in Brazil are being clear-cut as well as along the California and Oregon coasts and being reseeded with GMOs, all the same for-profit uni-trees. Trees are the lungs of our Earth that give off oxygen to feed humans, animals, and our oceans. They die, we die, its that simple.

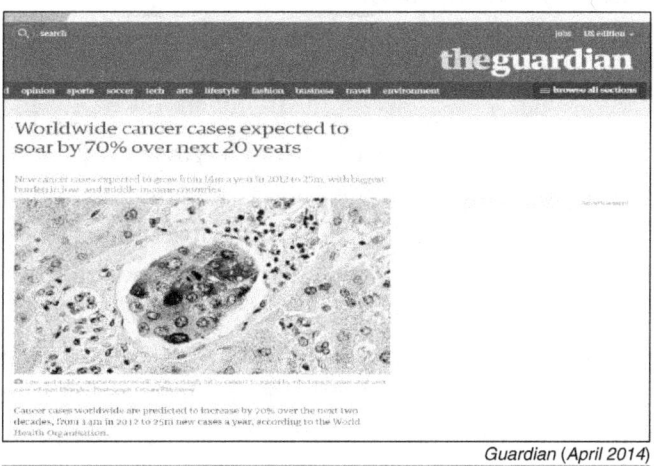

Guardian (April 2014)

"Levels of UVB are now often up to 1000% higher than official agencies are disclosing, these are extremely dangerous levels. How do we know levels are this high? Because we can and are metering UV radiation. We are now even detecting UVC radiation at the surface, UVC is the last band of UV radiation before x-ray radiation. We are told by all "official" monitoring agencies that UVC is stopped 100,000 feet up in the atmosphere, this is also a lie." — Dane Wigington, www.geoengineeringwatch.org

UVB: a band of ultraviolet radiation with wavelengths from 280-320 nanometers produced by the Sun. UVB is a kind of ultraviolet light from the sun (and sun lamps) that has several harmful effects. UVB is particularly effective at damaging DNA. It is a cause of melanoma and other types of skin cancer. It has also been linked to damage to some materials, crops, and marine organisms. The ozone layer protects the Earth against most UVB coming from the sun. It is always important to protect oneself against UVB, even in the absence of ozone depletion, by wearing hats, sunglasses, and sunscreen. However, these precautions will become more important as ozone depletion worsens.

Yet it is simple math, as long as no one is held accountable. Just like the TARP bailouts of over $2 trillion to the banksters that received from our government, (who used to work previously at the same banks). This is Wall Street at its best, privatizing profits and socializing costs while we, the "human resources", take the hit on our

Solar Dimming

Another side effect of large scale geoengineering practices is an artificial dome that aerosol spraying creates, which causes an unintended side effect of solar dimming. Due to man-made greenhouse effects, solar dimming reduces solar panel energy intake when Solar Radiation Management (SRM) operations are employed. Over the years I have recorded a more than 20% loss in solar energy uptake when sky painting is happening. The intensifying greenhouse heat inside the artificially created dome above us is another major side effect.

This is completely contrary and anti-ethical to the entire energy renewal movement and a big reason why solar panels are not mainstream. Less sunlight, less energy..

> *"Monsanto's GMO seeds are specially designed to grow in the high presence of aluminum. Aluminum is the chemical found in chemtrails. If this poisoning continues, true organic farming may become impossible in the not so distant future. When aluminum pollutes soil and water it kills crops. It collects in people and causes diseases!"* — Dane Wiggington

In 2011, the international Weather Channel, L.C. was purchased by Rockefeller Brothers, LLC. and nearly all social media devices now get their weather reports directly from the Weather Channel feed or from data supplied by some of the biggest military hardware sellers, Lockheed Martin Marietta and military think tank, Rand Corporation.

Here in California local and state weather reports are under reporting record high temperatures by some 5–10 degrees in the area where I live, by my own records. (Just like they have raised the acceptable radiation levels world-wide as well in recent years, due to the Fukishima radiation release in March 2011, greatly increasing man-made world radiation).

Chapter 6 | The Trees are Dying—We're Next

In 2014, measurements were taken and recorded by Dane Wiggingtion at www.geoengineeringwatch.org as to testing the Ultraviolet levels A & B. What he found through his own testing, was that UVB levels, much more deadly than UVA levels, were some 1000 % above official EPA reports. UVB is a much shorter, more intense burst of radiation and is the (bold) causal reason we all feel like the Sun is more intense on our skins. It is also the primary causal factor of the tree die-off we are seeing around the world today.

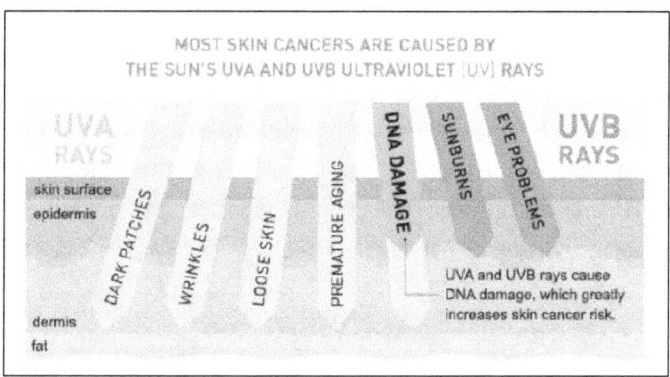

Increased UVB intensity causes a mass increase in skin, and other types of cancer, and dovetails, again so coincidentally, with reports in 2014 by the World Health Organization (WHO), who reported that cancer rates are predicted to increase by a whopping 70% by 2030, just fourteen years from now. Big Pharma and Wall Street look to profit handsomely from the die-off of human Life. Such is the capitalist system of the few profiting at the expense of the many and it is just "business as usual," until even those behind the levers of destruction will face the hell they knowingly have wrought on us. Remember, Nikola Tesla showed us free, non-polluting energy know-how back at the turn of the 20th century yet it has never been made available to the public. Damn the environment, profit we must.

It is important to understand that corporations derive from the word "corpse" and will do anything to profit, as long as none of the hierarchy are prosecuted and get frog-marched into jail. One shining example is Big Pharma, where over the past decade alone they have paid out over $31 billion in fines for practices that have disabled and cut short the lives of many with their drugs, made mostly in China

In 2013, Monsanto purchased The Climate Corporation for a cash price of $9,300,000. Monsanto's controlling the message makes for better manipulation of the farmers struggling from the geoengineered destruction of their crops and fields. Monsanto and Climate Corporation state how much they will be able to help the agriculture communities when in reality their goals have long since been clear. In Henry Kissinger's 1974 report, "National Security Study Memorandum 200" (NSSM 200), he directly targeted overseas food aid as an "instrument of national power." He has publicly declared the following: "Control the oil, you control the nation, control the food, you control the people." I learned another lesson very well from The Street, that once you get to the top you spend an inordinate amount of time and money to buy off politicians and energy to insure that you will be staying at the top, which is why the names Rothschild, Rockefeller, JP Morgan, etc., the Jesuits Court, Hofjuden's are still in control of our "current-sea" today.

Activism

It has gotten to such a sorry state that we are not allowed to label what is in our common foods any longer. Less than 1% of the country are farmers where the average age of farmers is 59 years young.

There is no longer any such thing as organic food, since all soil and water has been bombarded and contaminated through GE practices; this means we all have become synthetic humans in toto. We aware of GMOs in our foods but few are aware that since 2000, thousands of foods have been compromised with Bio-engineered, nano-foods at the sub-atomic level that enters easily through our blood and brain barriers.

Bio-engineered nano-foods are a more than $40 billion business that stays off the radar, while the FDA and USDA only play "advisory" roles without any sort of regulation or protection for the consumer.

As of Spring 2016, lawsuits are being filed in Federal and State court systems to halt geoengineering practices. Unfortunately, the courts are owned lock, stock, and barrel by the same powers that run the geoengineering programs but at this late stage in the game, it's do whatever we can, with whatever we have and start somewhere.

As noted previously, geoengineering our skies has been occurring for over half-a-century and most of us are just awakening to this very inconvenient truth today. Speaking out, being heard and helping to educate those that are unaware is the critical mission for all reading this book now.

Organizing community education seminars, taking soil samples to show heavy metal toxins, bringing them to the attention of your local air quality boards, and handing out DVDs that can be found at *www.geoengineeringwatch. org* are critical action items in the late, late stage of this, Humanities 6th Great Extinction Event.

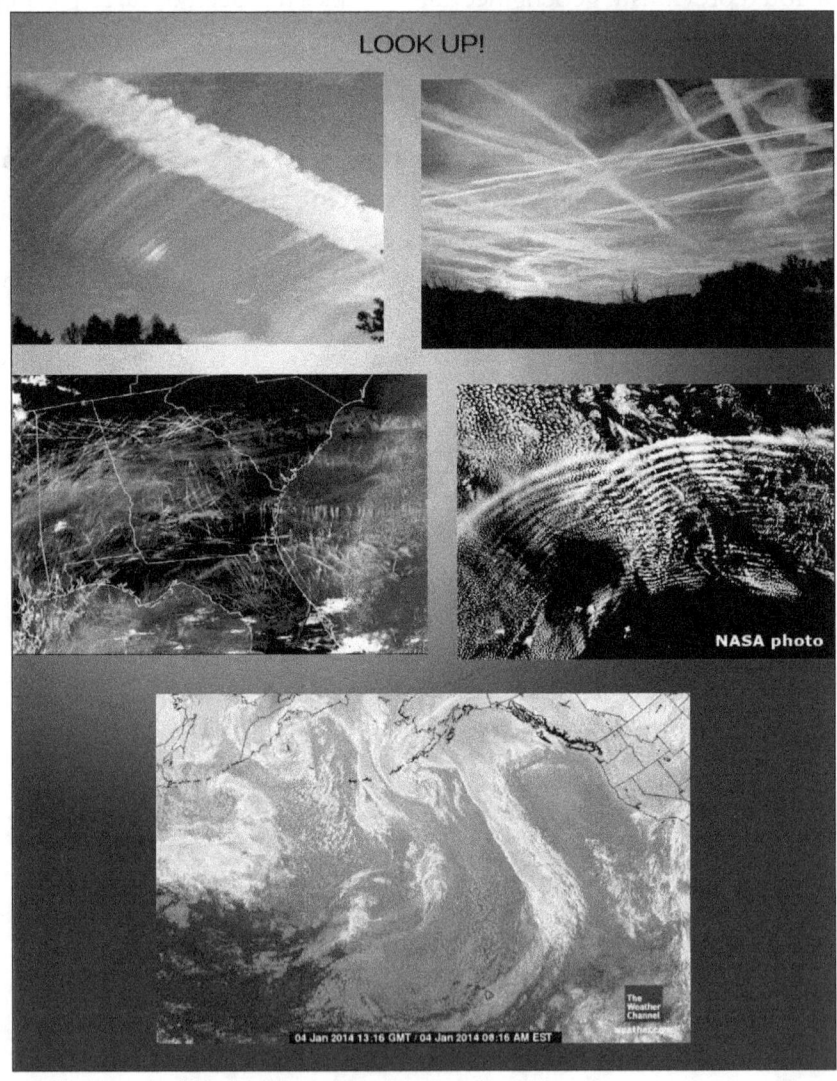

Chapter 7 | HAARP, LUCY, Alamo, and A.N.G.E.L.S. Projects: Advanced Technology Acceleration to Early Extinction?

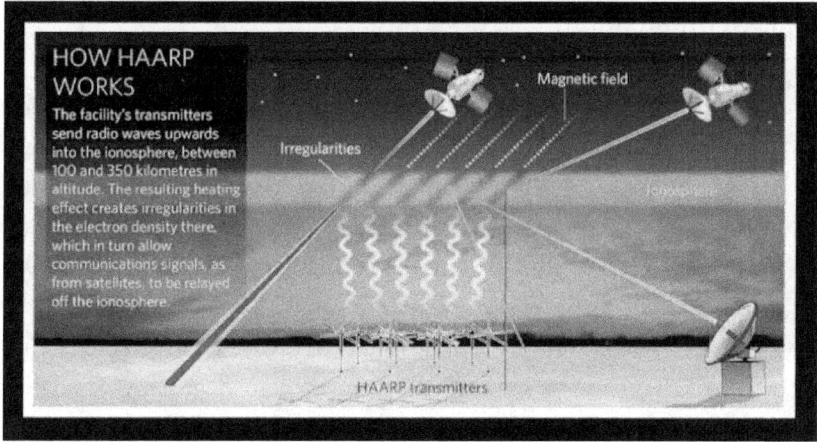

I think the reason I've occasionally said that is that it illustrates the kind of power that this technology grants us. And I think for better, for worse, what this technology gives us is this enormous kind of leverage and power to alter the climate and to do it with a very small amount of money or material and that power should frighten us, I think, and it presents real deep problems for governance. —Geo-engineering proponent, Harvard Professor David Keith

Methane Time Bomb—The Clatharate Gun

From Sam Arcana (a pseudo name for a group of scientists, so they can truth-tell and keep their jobs and pensions) at Arctic Methane Emergency Group (AMEG): "Huge quantities of methane are held in ice-like structures in the cold northern bogs and the bottom of the seas. They are called clathrates (or cathrates). They are stable only in the cold or under high pressure. Methane is 24x more potent a greenhouse gas than CO2."

The estimated amount of methane stored in these clathrates is gargantuan. It is the largest concentration of methane found on earth. The compression of methane gas in clathrates is enormous. One cubic meter of clathrates brought to the ocean's surface releases 164 cubic meters of methane.

The possibility of violent methane degassing (or "burping") has been called the clatharate gun hypothesis. There is a suggestion that the ocean's bottom waters couldn't warm up to 8°C. If so, that would certainly set off massive clatharate destabilization. This is what turns the clathrates into a ticking time bomb.

These hydrates are already being released with high altitude photos showing towering chimneys of methane bubbling off the ocean floor. They are subterranean versions of the gas field fires seen during the first Gulf War in Kuwait.

Historically there are spikes in the methane record that may be explained by the violent degassing of clathrates. Some think that the Ecocene hothouse period was caused by runaway global warming from clathrates released from the oceans."

"The biggest of these catastrophes occurred at the end of the Permian period some 250 million years ago. More than 94% of all marine species in the fossil records suddenly disappeared as oxygen levels plummeted and life teetered on the edge of extinction."

It took more than 20 million years for coral reefs to begin reestablishing and in some areas over 100 million years for ecosystems to reach their former healthy diversity. They were both were caused by temperature rises of less than 6.5°C (43.7°F). These temperatures are now average but in the Siberian permafrost where much of the clathrates are buried, the land is warming faster than anywhere else on Earth."

Chapter 7 | HAARP, LUCY, Alamo, and A.N.G.E.L.S. Projects

HAARP SBX Platform being towed out to sea and linked to Earth's geodesic energy grids for maximum connectivity.

The Three Major Methane Release Sources

A giant hole has formed in the hydroxyl-ozone layer in equatorial South East Asia and the West Pacific and allows methane sourced from the 9–11 km high leading edge of the southward migrating Arctic Atmospheric Global Warming Veil to rise unimpeded into the stratosphere where it is increasing in concentration.

The high equatorial concentration of methane will be the reason for the extreme "El Nino" this year and the heating extends eastwards to the Gulf Stream enhancing its energy in a giant feedback loop.

➡ A major feedback mechanism is the heat-trapping Methane Global Warming Veil sourced from the sub-sea Arctic methane hydrates, which has blanketed the entire United States as far south as the Gulf Coast (from July 2013, *www.methanetracker.org*). This atmospheric methane veil will further overheat the Gulf Stream thus returning even hotter water back to the Arctic sub-sea methane hydrate destabilization grounds generating more extensive methane eruption zones

➡ A second feedback mechanism is caused by the Arctic Ocean ice and Arctic region permafrost which is being degraded severely by atmospheric temperatures around 20°C above normal, especially in Siberia and Alaska. About 90% of the Arctic frozen methane lies in the top three meters which is thawing and the temperatures of the river water is rising assisted by high atmospheric temperatures caused by widespread Spring fires. This hot river water flows north into the Arctic ocean where it spills onto the East Siberian Arctic

Shelf (ESAS; Shakova etc., 2010, 2013) and into the Beaufort Sea, where it is destabilizing shallow methane hydrates releasing increasing quantities of methane directly into the atmosphere and increasing the concentration of the expanding Arctic Methane Global Warming Veil. Cenozoic pyroclastic volcanoes also occur on the west end of the ESAS, so destabilization of shallow methane hydrates is probably also opening deep seated, vertical fractures which will allow mantle methane to rise up into the atmosphere.

➡ A third major feedback mechanism is formed by a massive hydroxyl (and ozone) hole that has developed in the atmosphere above the western Pacific and Indonesia (Robert Scribbler, 2014). Hydroxyls are nature's air cleaners and they remove air pollution and methane from the atmosphere (Heicklen, 1976). The massive hydroxyl hole in Indonesia allows the southward spreading, 9–11 km high, Methane Global Warming Veil to rise up into the stratosphere where it then returns back at high altitudes to the northern Polar regions to further thicken and increase the warming of the globally spreading methane cloud.

➡ The massive hydroxyl hole in the atmosphere over Indonesia and the West Pacific also allows the shallow Methane Global Warming Veil to rise vertically into the dense equatorial stratospheric methane belt increasing its concentration. This is probably the source of the El Nino heat build up in the Pacific which is likely to occur in the Summer and Fall where the winds over the warmer ocean have shifted to the east."

Chapter 7 | HAARP, LUCY, Alamo, and A.N.G.E.L.S. Projects

HAARP facility in Alaska used to steer weather, increase and reduce hurricanes, earthquakes and frequency-hopping mind control.

If All You Have is a Hammer—HAARP

HAARP—the largest ionospheric heater in the world—capable of heating a 1,000 square kilometer (1,609 miles) area of the ionosphere to over 50,000 degrees. It's also a phased array, which means it's steer-able by HAARP technology and those waves can be directed to a selected target area. By sending radio frequency energy up into an artificially created ionospheric layers, geoengineers can and do manipulate storm weather and attempt to cool by focusing radio beams. Large areas of the Earth are being heated up.

The big answer to our major league weather problems is to stop all geoengineering practices today and let Nature heal herself.

The evidence is overwhelming, abundant, and clear for those with eyes to see and ear to hear that the geoengineering of our skies for the past five decades and more, is accelerating the rapid die-off of all sentient life on Earth. Since all western science is based on a big hammer to pound and break apart, they have created a High Frequency Active Auroral Research Program (HAARP) program to break up the large methane release pools in the Arctic regions through frequency manipulations of our skies.

It is like an opera singer who can break glass with her high-pitched voice. Geoengineering creates a false ionospheric sky to bounce signals off, sometimes as low as 10–15,000 feet above our heads.

The theory recently developed in science labs and being put into practice is to use alternating frequencies from HAARP devices around the world to break up methane gases as they are released from the Arctic Tundra and before they can form life-killing CO_2. These nano-particulate matters then create what science calls "nano-diamonds;" very small methane crystals that they hope will cause large-scale methane to spread around the world for decades to come. The name for one of their projects (explained below) has been named "LUCY," as in the Beatles song, *Lucy in the Sky with Diamonds!*" I wish I was making this up but I am not.

Geoengineers are using highly advanced technology experimented in government laboratories to blast the large pool of methane gases using HAARP technology.

Climate Weather Engineering

Tragically for all, by all measures, they have been wildly unsuccessful in creating snow to re-Arctify and keep the methane pools from releasing and causing even more rapid heating of Earth. Over the past few years a vicious warming feedback loop has occurred with unintended, yet predictable consequences. In 2007, the first large methane pools released as the snow and ice melted and exposed the tundra. Then much larger pools of methane gas released in 2013. Today, in the Spring of 2016, the lid is off and we'll likely be ice free this Fall for the first time in millions of years; a very, very big deal for the future of all life on Earth.

> **K**eep in mind that DARPA, the US Defense Advanced Research Projects Agency has advanced technology some 25–100 years beyond what they tell us today. It was DARPA that released the Internet platform in 1995 that we use today. (The Internet speed we are used to today is the same speed DARPA had in 1995) They also hold the Nikola Tesla patents for free-energy developed by Tesla in the early 1900s but held it back from the public still to this today.

Yet, the spin doctors of corporate science, hiding such critical news of the increasingly deadly, large-scale methane releases, only want to point to increased snow in the Arctic, taken from computer generated images from Google Earth. These images only show a few inches of increased snow coverage after snow nucleated spraying, whereas, previous snow cover was hundreds of feet thick in some areas, not inches, as shown currently. (*Note that we see only CGI images of the Arctic and Antarctic Circle, never real-time live pictures from NOAA and NASA*).

The artificial heating literally lifts the ionosphere within a 30 mile diameter area, therein changing localized pressure systems and perhaps the jet stream paths. Moving a jet stream is a phenomenal feat, that man is capable of doing this, there is no precedent for this arrogant attempt to play God with the world's weather. This goes on under the radar of the "sleeping sheeple" who are clueless to the extreme peril we currently face, while our government plausibly denies such operations are taking place. Yet, as you will see below, highly advanced technology is being employed in desperation to save us from our own extinction.

Factoid: The word "World" is derived from ancient Phoenician language from the word "Whirled," as in a spiraling, imploding toroidal vortex which occurs regularly at the North Pole.

The Weapons of Mass Weather Manipulation (WMWM) was brought into this world over a hundred years ago by Nikola Tesla and patented by Bernard Eastlund. (US PATENT 4,686,605). Bernard Eastlund's discoveries are innovative applications of the work of Tesla and Michael Faraday. Tesla's plans to provide power on any spot on the Earth, to modify weather and eliminate drought, floods, and hurricanes was usurped by the US military to make a death ray.

Cosmic Orgone Engineering (CORE) and Wilhelm Reich

Wilhelm Reich (March 24, 1897–November 3, 1957) was an Austrian-American psychiatrist and psychoanalyst, known as one of the most radical figures in the history of psychiatry

Mr. Reich designed a "cloudbuster;" he said he could manipulate streams of orgone energy in the atmosphere to induce rain by forcing clouds to form and disperse. It was a set of hollow metal pipes and cables inserted into water, which Reich argued could create a stronger orgone energy field than was in the atmosphere, the water drew the atmospheric orgone through the pipes.

The cloudbuster was intended to be used like a lightning rod—focusing it on a location in the sky and grounding it in some material that was presumed to absorb orgone—such as a body of water—it would draw the orgone energy out of the atmosphere causing the formation of clouds and rain.

Reich conducted dozens of experiments with the cloudbuster, calling the research "Cosmic Orgone Engineering." The official US Government position on "Orgone Energy" is that it doesn't exist. Yet, that didn't stop the US Government from passing laws against it, burning all of Wilhelm Reich's books, destroying all his equipment, and locking him up in prison (where he died) for something they said, *"Doesn't exist."*

Today the world military has the abilities to command and control all weather as well as cause earthquakes. They can steer hurricanes, create snow and alter temperatures at will. Case in point was the Beijing Olympics, where reports of over a billion dollars in damages was created from the

Chapter 7 | HAARP, LUCY, Alamo, and A.N.G.E.L.S. Projects

admitted and acknowledged use of weather modification technologies applying HAARP technology was used.

HAARP has also been reported to have been highly active off the coast of Fukishima in March of 2011 during the earthquake and resulting tsunami that caused nuclear reactors to fail and release massive toxic radiation into the air and sea that continue to this day.

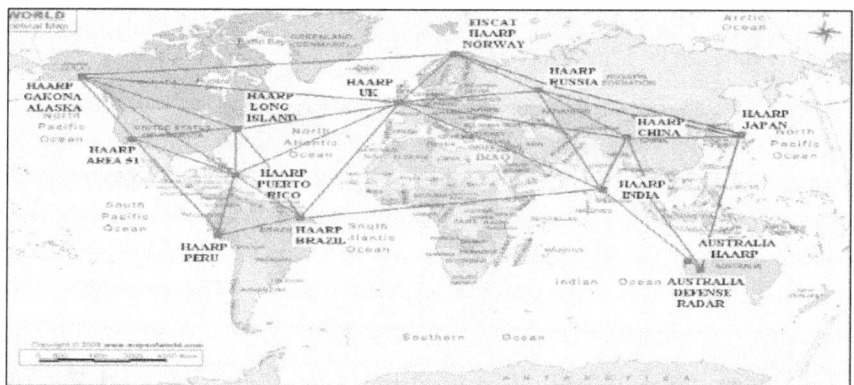

HAARP is a military project based on Eastlund's discoveries. It has eluded the spotlight of the popular press. But for years now, conspiracy and coincidence theory web sites as well as some credible scientific publications have questioned how playing God with the ionosphere could be hurting the environment.

To make matters worse, HAARP has been maintained as a mostly clandestine project operated by the US Navy, world-wide and coordinated through "The Entity," the intelligence gathering arm of the Vatican that lords over the CIA, NSA, MI5 and 6, and Israeli Mossad spy spook agencies..

Information that is made available to the public is carefully worded to make HAARP appear as a bland, harmless, unclassified, atmospheric research facility. Recently, the US Government announced the shuttering of the HAARP facility in Alaska, yet as you can clearly see from the above map, HAARP is situated world-wide along Earths geodesic energy grid key lines, stationed all around the Antarctic Circle, and over energy Earth energy power grid line in our ocean.

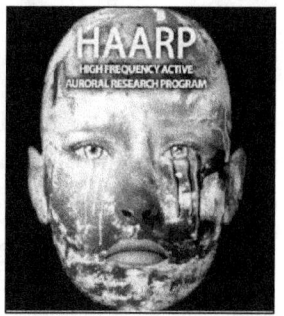

The Earth's electromagnetic energy grid lines also pass through ancient historic monuments like Stone Henge, the Egyptian pyramids at Giza, Mexico, Bosnia, the Georgia Guidestones and undersea in Japan as well as many prominent Christian churches that built their monuments directly over ancient pagan sites. Our ancient ancestors knew the power of Earth's energy fields and so today, does our overlords of technology use the same energy power centers to control and command weather today.

It has also been proven that Earth has its own electrical magnetic frequency (8.47 and 8.54 MHz), which is nearly the same frequency as Earth's ionosphere. It is exactly in the same range as the vibrational frequency of human beings.

Factoid: The powers in charge enlisted the help of shoe manufactures in  the 1950s and 60s to replace leather soles on shoes with rubber. This simple act disconnected us from the grounding source to our Mother Gaia, and thus our own toroidal connection to the heavens above. Since we are all electromagnetic beings and rely on a positive earth charge to connect to our own bodies battery, this simple act has denied many people of accessing their own higher consciousness. It was celebrated, for those in the know, with the Beatles album cover in 1965, "Rubber Soul." On a related note, most of the Beatles, Rolling Stones, David Bowie and other "British Invaders" music was written by Theodore Adorno and his cohorts at the Tavistock Institute.

Chapter 7 | HAARP, LUCY, Alamo, and A.N.G.E.L.S. Projects

Some Interesting "Then Year" BW Possibilities

- Aflatoxin - ("natural," parts-per-billion, carcinogen)
- Airborne varieties of Ebola, Lassa, etc.
- Binary agents distributed via imported products (Vitamins, Clothing, Food)
- Genomicaly (individual/societal) targeted pathogens
- Long term/fingerprintless campaign (as opposed to "shock and awe" BW)

An (Existing Bio Calmative - VEE (Venezuelan Equine Encephalitis)

- **Ideal Incap. BW Agent**
- **Weaponized by U.S. & USSR in 50's/60's**
- **Easily transmitted via Aerosol**
- **Highly infectious, Low Fatality Rate**
- **1 to 5 day incubation, 3 week recovery**
- **Tested on Humans (Operation Whitecoat)**
- **No Treatment Available**

"Man is the only species that is willfully and knowingly destroying the only home that gives all Life" —*Extinction-R-Us Through Advanced Technology?*

Most of the information below further explores and shows new proof that Humanity is facing an Arctic methane induced firestorm withing the next one or two decades. Adding to the great warming, methane emissions from Arctic sub-sea methane hydrates and is now out of control most notably along the eastern and western coasts of the American continent and Asia so far.

LUCY, ALAMO and the A.N.G.E.L.S. Projects
(Taken directly from Sam Carana's Arctic-news blog and linked in the bibliography and resources section).

"Massive temperature increase is almost entirely due to the build up of methane in the atmosphere from destabilization of the sub-sea Arctic methane hydrates. The Arctic atmosphere methane is the killer and we need to immediately find methods to remove large volumes of it from the troposphere and the stratosphere if we have any hope of surviving the fast approaching extinction event. What we have got to do is eliminate as much of the atmospheric methane by whatever means we are able to devise, to bring its concentration down to about 700 PPB. This level will eliminate much of the methane delayed temperature anomaly and give the massive industrial nations a little leeway to get their houses in order.

Chapter 7 | HAARP, LUCY, Alamo, and A.N.G.E.L.S. Projects

All the scientific expenditure and ingenuity of the major industrial nations should be engaged in developing methods of breaking down atmospheric methane without burning it. Methods of increasing the tropospheric and stratospheric hydroxyl concentrations and using radio – laser systems such as the Alamo – LUCY projects and their applications to HAARP must be developed and tested with the utmost urgency, as should local methods of converting carbon dioxide and methane via catalysts into other products.

We have to get rid of this methane monster before it devours us all. If we fail to reduce the fast growing methane content of the atmosphere in the next few decades we are going to go the same way as the dinosaurs."

The LUCY Project

The LUCY Project is a radio beat frequency and laser system for destroying the first hydrogen bond in atmospheric methane when it forms dangerously thick, global warming clouds over the Arctic. It generates similar gas products to those normally produced by the natural destruction of methane in the atmosphere over some 15–20 years. This system will use similar frequencies to those used in generating nano-diamonds from methane gas in commercial applications over the entire pressure range of the atmosphere from the surface to 50 km (31 miles).

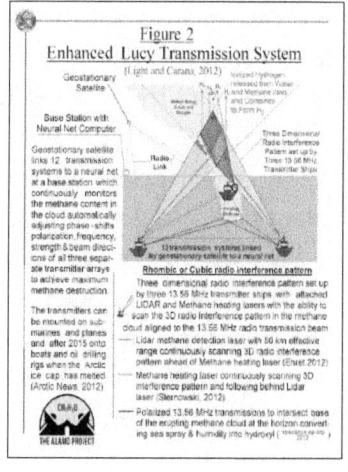

Methane produced at the surface diffuses upward and is broken down by photo dissociation (sunlight) and chemical attack by nascent oxygen and hydroxyl. Methane destruction with nano-diamond formation! Modified LUCY Project to Generate Hydroxyls at the Sea Surface using Beams of Polarized 13.56 MHz Radio Transmissions. The LUCY and Alamo (HAARP) projects were designed to break down atmospheric methane using radio–laser transmissions.

In this system three additional transmitters on three separate ships will have their antenna placed slightly lower than the main 13.56 MHz methane destruction antennae. Recent experiments have shown that when a test tube of seawater was illuminated by a polarized 13.56 MHz radio beam, that flammable gases (nascent hydrogen and hydroxyls) were released at the top of the tube.

In the modified version of the LUCY Project, hydroxyls will be generated by a polarized 13.56 MHz beam intersecting the sea surface over the region where a massive methane torch (plumes) is entering the atmosphere in order that the additional hydroxyl produced will react with the rising methane breaking a large part of it down. In the Arctic Ocean, the polarized 13.56 MHz radio waves will decompose atmospheric humidity, mist, fog, ocean spray and the surface of the waves themselves into nascent hydrogen and hydroxyl."

A newly determined global atmospheric temperature gradient indicates that the mean global atmospheric temperature anomaly will reach 1.5°C in 15 years (2028.5) and 2°C in 20 years (2033.4). Consequently we only have 15 years to get an efficient methane destruction radio – laser system designed, tested and installed (LUCY and Alamo (HAARP) Projects) before the accelerating methane eruptions take us into uncontrollable runaway global warming.

This will give a leeway of 5 years before the critical 2°C temperature anomaly will have been exceeded and we will be looking at catastrophic storm systems, a fast rate of sea level rise and coastal zone flooding with its disastrous effects on world populations and global stability. An anomalous temperature of 4°C will be reached by the atmosphere around 2043 which will end the vegetation carbon sink, preventing plants from helping balance carbon dioxide exhalation and this will further accelerate climatic change."

Alamo Project

"Methane is rising into the stratosphere and mesosphere where some of it is being oxidised to produce larger quantities of noctilucent clouds between 76 and 85 km altitude (solar dimming). Noctilucent clouds were originally confined to the southern polar regions, were then seen north of Norway and are now occurring at much lower latitudes over Colorado. Professor James Russel of Hampton University argues that the build up of methane in the atmosphere is the reason for the increase in noctilucent clouds. Prof Russel says: "When methane makes its way into the upper atmosphere it is oxidised by a complex series of reactions to form water vapour. This extra water vapour is then available to grow ice crystals for noctilucent clouds."

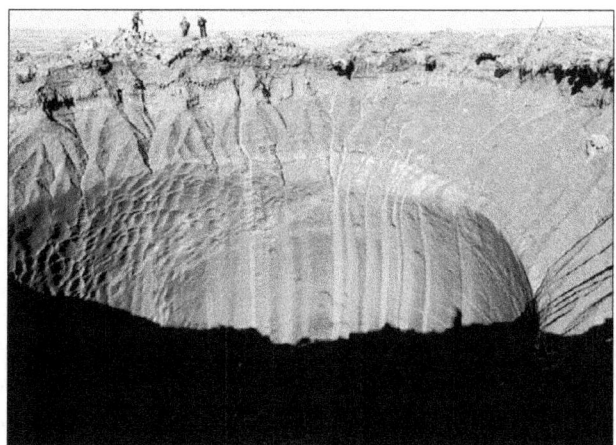

Huge Methane Blow Hole (2013) Serbia, Russia (note people on rim for size perspective)

Massive Methane Release hole, Siberia, USSR, 2013

If we succeed in breaking down the methane in the stratosphere and mesosphere with the HAARP–IRIS (Ionospheric Research Instrument) using the 13.56 MHz methane destruction frequency, it could lead to an increase in noctilucent cloud formation in a circular zone directly above the HAARP transmitters which could be detected by optical cameras or radar. Besides the elimination of the high global warming potential methane, noctilucent clouds formed from methane water condensing on meteorite dust and nano- diamonds will reflect the suns radiation back into space and this will also help to counteract global warming. The HAARP-IRIS transmitters normal frequency range is from 2.8 MHz to 10 MHz. If for example a 10 MHz carrier wave is modulated by a 3.56 MHz signal, it will produce and Upper Side frequency of 13.56 MHz, the necessary methane destruction frequency and a Lower Side Frequency of 6.44 MHz (Penguin Dictionary of Physics, 2000).

The HAARP tests should be conducted in the summer when the stratospheric temperatures are at the lowest in Alaska (140°K–160°K) increasing the chances of noctilucent cloud formation from the radio frequency oxidised methane."

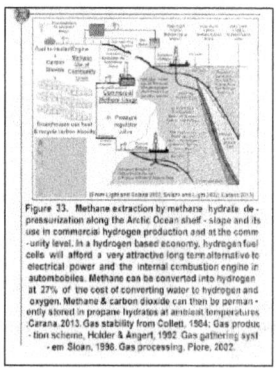

Figure 33. Methane extraction by methane hydrate de-pressurization along the Arctic Ocean shelf - slope and its use in commercial hydrogen production and at the community level. In a hydrogen based economy, hydrogen fuel cells will afford a very attractive long term alternative to electrical power and the internal combustion engine in automobiles. Methane can be converted into hydrogen at 27% of the cost of converting water to hydrogen and oxygen. Methane & carbon dioxide can then be permanently stored in propane hydrates at ambient temperatures. Carana. 2013. Gas stability from Collett, 1994; Gas production scheme, Holder & Angert, 1982; Gas gathering system Sloan, 1998; Gas processing, Pioro, 2002.

"We clearly have to destroy more than 62% to 64 % of the present methane content of the atmosphere before the Earth will have a livable atmosphere. This also means stopping all the Arctic methane eruptions by de-pressurizing the methane under the sub-sea methane hydrates to stabilize the methane content of the atmosphere." —Sam Carana 2013

A.N.G.E.L.S. Project
(Arctic Natural Gas Extraction, Liquefaction, Sale and Storage)

"The A.N.G.E.L.S. Projects are solutions to the extreme Arctic methane build-up that must be done in conjunction with a complete cutback in carbon dioxide emissions from North America, if we have any hope of stopping this now, almost out of control, exponentially escalating build-up of methane in the atmosphere and its extreme enhancement of global warming.

The A.N.G.E.L.S. Project will require the drilling of inclined boreholes into the sub-sea Arctic permafrost/methane hydrates to drain the over pressured methane from beneath them and thus de-pressurize the reservoirs.

This will stop many of the Arctic sea surface methane eruptions by drawing ocean water down the eruption zones and will allow a controlled destabilization of the undersea methane hydrates producing natural gas source for many hundreds of years. This gas can be liquefied in surface plants, put into lNG tankers and sold as a feedstock for hydrogen plants or permanently stored in propane-ethane hydrates with carbon dioxide at ambient temperatures in deep ocean basins.

Symbiotic Bacteria Destroying Methane in the Arctic Ocean

Two symbiotic methane eating microbes in cold ocean waters excrete carbon dioxide and need an enzyme that requires tungsten to operate. They are a sulfate utilizing *deltaproteobacteria* and an anaerobic *methanotrophic archea*. The carbon dioxide released by the bacteria reacts with minerals in the water to form calcium carbonate. If these bacteria and tungsten enzymes were introduced in great quantities into the Arctic Ocean they could eliminate the methane plumes before they entered the atmosphere giving humanity time to destroy the existing atmospheric methane accumulations."

It needs to be noted that all of this advanced technology is being tested on Earth itself as a grand lab experiment. Like the cancer, that never ever gets cured, only made worse by our ever toxifying environment, so far LUCY and Alamo Projects have made a dire situation now considerably worse by all measurements Accelerated. These geoengineering practices have caused deadly greenhouse gas accumulations all over the world, which in turn inversely accelerates Earth heating up, truth be told.

Chapter 8 | **Conclusion**

(*From Sam Carana www.amegblog.me, (2013)*)

"The Earth is a giant convecting closed system, the underlying molten magma being heated by deep seated radioactivity and the oceans and atmosphere are its cooling radiator which allows the Earth the facility to vent this heat into open space. Mother Earth has carefully held the atmospheric temperature within a stable range necessary for oceans to exist for at least 4 billion years and nurtured the earliest bacteria to evolve into today's space faring humans.

The fouling up of the Earth's cooling radiator from human emissions of greenhouse gases derived from fossil fuels will be counteracted by Mother Earth in her characteristic fashion by emitting vast volumes of deadly methane into the atmosphere from the Arctic regions.

This will lead to the total extermination of all harmful biological species that produce greenhouse gases in the same way that Mother Earth did during the Permian and other extinction events.

In this case however we have totally tipped the balance with our extreme carbon dioxide and methane emissions so that there will be no chance of recovery for the Earth in this time frame, because the methane release will cause the oceans to begin boiling off between 115°C and 120°C (239°F–248°F) in 2080 and the Earth's atmosphere will have reached temperatures equivalent to those on Venus by 2096, 460°C–467°C (869°F–872°F). This is also being coined as the Venus Effect.

Mankind's greed for fossil fuels will have completely destroyed a magnificent beautiful blue planet and converted its atmosphere into a barren, stiflingly hot , carbon dioxide rich haze. The earth will have moved permanently out of the magical zone (Circumstellar habitable zone, Goldilocks zone) where life (some of it probably highly intelligent) also exists elsewhere in the myriad of other solar systems that are located within the far reaches of our Universe.

The power, prestige and massive economy of the United States has been built on cheap and abundant fossil fuels and Canada is now trying to do the same. The present end of the financial crisis and recovery of the US economy will take us down the same fossil fuel driven road to catastrophe that the US has followed before. Unless the United States, Canada reduce their extreme carbon footprints (per unit population), they will end up being found guilty of ecocide and genocide as the number of countries destroyed by the catastrophic weather systems continues to increase."

"The United States and Canada with their expanding economies and their growing frenetic extraction of fossil fuels, using the most environmentally destructive methods possible (fracking and shale oil) as well as the population's total addiction to inefficient gas transport is leading our planet into suicide. We are like maniacal lemmings leaping to their deaths over a global warming cliff. What a final and futile legacy it will be for the leader of the free world to be remembered only in the log of some passing alien ship recording the loss of the Earth's atmosphere and hydrosphere after 2080 due to human greed and absolute energy ineptitude.

Chapter 8 | Conclusion

The US has to put itself on a war footing, recall its entire military forces and set them to work on the massive change over to renewable energy that the country needs to undertake, if it wishes to survive the fast approaching catastrophe. The enemy now is Mother Nature who has infinite power at her disposal and intends to take no prisoners in this very short, absolutely brutal, 30–40 year war she has begun. I cannot emphasize more, how serious humanity's predicament is and what we should try to do to prevent our certain final destruction and extinction in the next 30–40 years if we continue down the present path we are following."

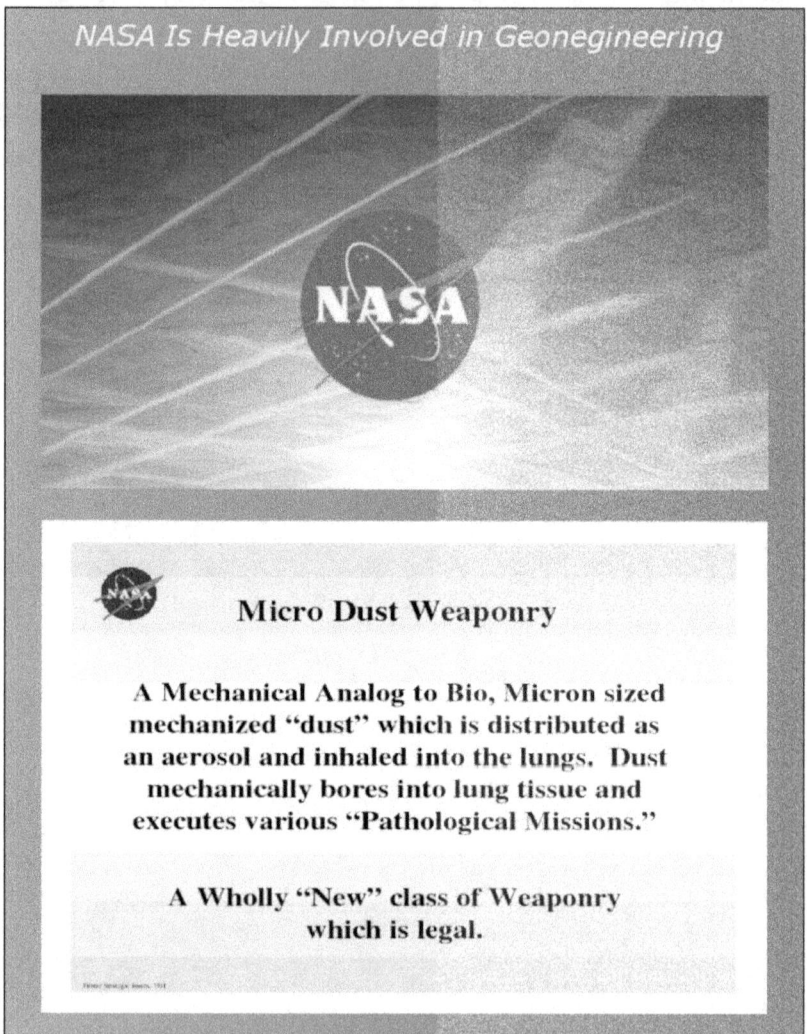

Chapter 9 | Summary

Speaking the truth of our anguish for the world brings down the walls between us, drawing us into deep solidarity. That solidarity is all the more real for the uncertainty we face. —Joanna Macy

Clearly we have caused the Earths natural weather systems to be greatly altered by anthropogenic means, so the answer is stop it. Stop it all and begin a massive conservation effort at the grass-roots level using what our communities have available, where you live.

NASA, Not a Civilian Space Agency

NASA and other alleged space agencies have been heavily involved in conducting the subversive spraying of our skies for decades now. They have even documented their goals and future plans for their Full Spectrum Dominance of Earth in coordination with the world's military power structure of today. NASA is funded under the Department of Defense and as a military operation was founded in 1958 by Nazi rocket scientists, movie producers, and Freemasons, yet has always been promoted only as a civilian operation.

The screen shots above show the NASA document, *"Future Strategic Issues, Future Warfare,"* written around 2005. The PDF is listed in the Website and Resources section of this book. NASA has done extensive research and development of many different aerosol application techniques and uses including very disturbing work with self-replication nanobots applied through aerosol application of *"Smart Dust,"* which we are all breathing, as they spray over our heads on a regular basis, which I cover in Book III, *"Touchless Torture. Target Humanity."*

> *Make no mistake, NASA is a primarily a military operation as evidenced by there own slide show presentation authored by their Chief Scientist at NASA Langley Research Center, —Dennis M. Bushnell.*

NASA's very first Director of the Saturn Rocket Program was Werhner Magnus Maximilian, Freiherr von Braun who held membership No.5,738,692 with the Nazi Party of Germany when he directed the design of V1 and V2 rockets ("*V*" for Vengeance) that  rained hell-fire on US allies during WWII, killing tens of thousands.

Von Braun, along some 3000 other Nazi German scientists, including some of the most heinous eugenicists and mind control specialists, were brought into the US over three decades, beginning in the early 1950s under Operation Paperclip. (*Paper clips were used on their files as code for the operation.*) The operation was run under the Office of Strategic Services (OSS), now the CIA, and conducted by the Joint Intelligence Objectives Agency (JIOA).

Another Nazi scientist who became an American space hero was Dr. Hubertus Strughold, later called "the Father of US Space Medicine." He had a long and distinguished career at NASA and even had an American library named after him. He ran a facility at Dachau in which medical experiments were carried out on prisoners and he even had a mobile laboratory traveling from camp to camp. One aspect of Mr. Strughold's research was a precursor to the CIA's MKUltra mind control experiments, which today has now become a wireless touch-less torture program using geoengineering, HAARP, super-computers, Smart Dust, and some very, very sick minds that only desire to command and control.

Mr. von Braun also found time in 1958 to host the Disney show, *"Man in Space,"* to start promoting the US Space Program. His co-host was Heinz Haber, another NASA Nazi. Mr. Haber worked for Mr. Strughold and co-authored papers that were based on human experiments performed at Dachau and other concentration camps in which hundreds of prisoners were subjected to experiments that simulated the conditions of high speed, high altitude flight. Prisoners that survived the experiments were generally killed, then dissected.

When the Eisenhower administration asked Walt Disney to produce a propaganda film regarding the peaceful uses of nuclear energy, Mr. Haber was picked to host the Disney show, *"Our Friend the Atom."* Mr. Haber then wrote a popular children's book of the same title.

The JIOA worked independently to create false employment and political biographies for the scientists. The JIOA also expunged from the public record the scientists' Nazi Party memberships and regime affiliations. Once *"bleached"* of their Nazism, the scientists were granted security clearances by the US government to work in the United States and given US Passports.

A well-hidden fact of revisionist history is that before and during the entirety of WWII many of the greatest US corporations were partnered with Germany.

John D. Rockefeller provided the anti-knock fuel additives through his Standard Oil of New Jersey so Germany's Luftwaffe Air Force could fly. Germany has never been an oil producing country and had to manufacture synthetic fuel for the war effort. Without no-knock fuel additive no airplanes would be able to fly for the Nazis. Of course these facts are hidden in plain sight and never taught in our public indoctrinated school systems.

The German behemoth chemical company, I.G. Farben, is known to have provided the military with chemicals used on Allies to torture and gas war prisoners.

On I.G. Farben's US subsidiary board of directors, up to and through WWII, were some of the most prestigious names among American industrialists such as: Edsel Ford of the Ford Motor Company, C.E. Mitchell, and Walter Teagle, directors at the Federal Reserve Bank and President Franklin D. Roosevelt's Georgia Warm Springs Foundation.

According to Antony Sutton, in his brilliant trilogy of books on Wall Street and Stalin, Hitler and FDR; provided documentation showing that I.G. Farbern and other businesses of the world elite were spared allied bombing throughout WWII.

Paul M. Warburg, the first director of the Federal Reserve Bank of New York and chairman of the Bank of Manhattan (later Chase Bank), was also an I.G. Farben director.

Meanwhile, in Germany, his brother, Max Warburg, was also a director at I.G. Farben. Max Warburg belonged to a circle of advisors of the German emperor, Wilhelm II. Following World War I, Max Warburg participated in the negotiation of the Versailles Treaty as member of the German delegation, though he declined the offer to chair the Finance committee and instead suggested his partner in the Warburg bank, Carl Melchior.

Paul Warburg married Nina Loeb, daughter of Salomon Loeb, a founding partner of the New Yorker Wall Street banking house Kuhn, Loeb and Co. Felix Warburg, another brother married Frieda Schiff, the only daughter of Jacob H. Schiff, a senior partner in the same firm. Both Warburg brothers eventually joined as partners of Kuhn, Loeb, the second biggest private bank in the United States prior to WWI.

Prescott Bush, father and grandfather to two US Presidents, was indicted under the Trading with the Enemies Act in 1942 by the United States for his continued illegal and treasonous assistance to the Germans during wartime. His assets were seized and he likely promised his fellow cronies of capitalism never, ever to do it again.

Another great omission in the US history narrative; car maker Henry Ford, received the Grand Cross of the German Eagle for his support to Germany. He also opened in the 1930s the world's largest automobile manufacturing plant in Gorsky, Russia. The plant was later retrofitted to make armored tanks used against the US in the Korean and Vietnam wars. When WWII broke out he abandoned some 450 US auto workers and their families to help run the new Russian factory and is recounted in the excellent book by Karl Tobien, *"Dancing with the Red Star,"* by the only known survivor of the Russian gulags, Margaret Werner.

Coca Cola changed their brand name to Fanta, to continue to sell sodas in Germany. IBM contributed new computer know how to keep track of POWs and extermination records. The list goes on to include many

US truck parts manufactures and medical testing device companies and exporting scientists to the US with complete immunity from prosecution for their war crimes.

> *"You can't solve a problem from the same mind that created the problem in the first place."* —Albert Einstein

It's completely naive to believe that the same ones causing the problems are going to suddenly wake up and find the error of their ways. It will only be by the grace of the Great Mystery that we will be allowed to survive the near future. Hope comes through active consciousness awakening and education resulting in A-C-T-I-O-N. Current projections from those in the know believe we have less than a decade left for most sentient life on Earth if we do not stop all geoengineering activities today as well as greatly reduce CO_2 polluting activities.

We are the Problem. Yet, climate engineers playing God with our weather are making it exponentially worse, while spraying heavy metal toxins over our waterways and on our soil we where our food is grown. The Baby "Doomers" of my generation choose to remain absolutely clueless as to the effects of Earth's warming liquidation.

All food now shows ever increasing levels of aluminum and other nano-particulate matter. Because of the exposure to heavy metal toxins we have all become antennas that have the proven ability to manipulate anyone at will.

Much sooner than later, foods will only be able to be grown with Monsanto related heavy-metal resistant seeds. Growing your own food, and or partnering with local farmers is a must as well as using heavy metal lifting plants for mitigation of the soil like sunflowers and hemp plants.

Let's be very clear here, Mother Gaia will survive humanity's suicidal self-genocide, yet those that rely on Earth for their subsistence and survival will not...and this is not a big IF, it is now a *"when"* because we have passed the *"we had better soon or else"* moment in our collective history. We are past a normalized world, yet our only hope of survival is if we allow our Supreme Mother and Father Sky to heal and come into balance once more

All hands on deck are called to action immediately to sound the alarm to awaken the snoring. Inaction is deemed more deadly than error at this late hour, the 6th Great Extinction Event in the History of Mankind. If we fail—the message is clear—our children and all Life will have no long term future. And that is one hell of a legacy to leave on or way out.

Truth be told.

The End

Other Author Books
(to be released in 2016)

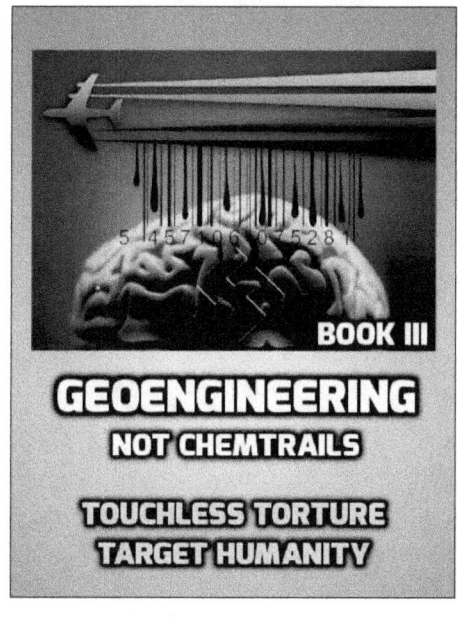

Epilogue

A Hopi Elder Speaks:
"You have been telling the people that this is the eleventh hour,
now you must go back and tell the people that this is the hour.
And there are things to be considered ....

Where are you living?
What are you doing?
What are your relationships?
Are you in right relation?
Where is your water?
Know your garden.
It is time to speak your Truth.
Create your community.
Be good to each other.
And do not look outside yourself for the leader."

Then he clasped his hands together, smiled, and said,
"This could be a good time!"
"There is a river flowing now very fast. It is so great and swift that there are
those who will be afraid. They will try to hold on to the shore. They will feel they
are torn apart and will suffer greatly.. Know the river has its destination."

The elders say we must let go of the shore, push off into
The middle of the river, keep our eyes open, and our
Heads above water. And I say, see who is in there with
You and celebrate. At this time in history, we are to take
Nothing personally, Least of all ourselves. For the moment
That we do, our spiritual growth and journey comes to a halt.
The time for the lone wolf is over. Gather your selves!
Banish the word struggle from your attitude and your vocabulary.
All that we do now must be done in a sacred manner.
We are the ones we've been waiting for.

Appendix I

Harvard Professor and Lead Government Scientist, David Keith, Promotes Aerosol Geoengineering as Fix for Warming via Covert Climate Modification

This interview reveals a few interesting forecasts:

- Keith sees geoengineering (GE) programs to begin in no less than 10 years.
- Sulfur has been ruled-out as a GE agent but we know he recommended aluminum in GE activities at the 2010 AAAS conference but—for some reason did not mention aluminum in this interview.
- Keith claims the topic of GE has been "taboo" even though citizens have been complaining about it since the mid 1990s.
- Keith boldly claims the GE operation will be "invisible." If aerosol GE begins in 10 to 15 years, the population will already have adapted to seeing the sky littered with aerosol clouds thus nothing will have "appeared" to change in the perception to most people looking up.
- Keith invokes the guilt message that CO_2 is the by-product of people using too many resources and that the consumers must pay for the sins of corporations who caused the problem in the first place. His emphasis is never on the greed aspect of the so-called "elite."

Interviewer and host Tony Jones, appears clueless on the issue of covert aerosol operations.

✦ ✦ ✦ ✦

•Interviews on the following pages have been reproduced including bold emphases in its entirety from *www.chemtrailplanet.com*

David Keith Promotes Aerosol Geoengineering as Fix for Warming via Covert Climate Modification

TONY JONES, presenter:
"Earlier today I spoke with geoengineering expert David Keith, Professor of Applied Physics at the Harvard School of Engineering and Applied Sciences." (He was in Calgary, Canada). "David Keith, thanks for joining us."

DAVID KEITH, Applied Physics and Environmental Engineering, Harvard: "Great to be here."

TONY JONES: "Now scientists originally calculated that the major impact of global warming would happen towards the end of this century, so geoengineering was considered to be something far off in the distant and really science fiction for most people. Why the urgency now? Why has the debate changed?"

DAVID KEITH: "I think the debate's changed really because the sort of taboo that we wouldn't talk about it has been broken. So, people have actually known you could do these things for better or for worse for decades, actually since the '60s, but people were sort of afraid to talk about them in polite company for fear that just talking about it would let people off the hook so they wouldn't cut emissions."

"And that fear was broke a few years ago and so now kind of all the research is pouring out really because effectively had been suppressed, not by some terrible suppressor, but by a fear of talking about it."

TONY JONES: "So what do you think would actually drive the world's superpowers or a collective of nations to decide to actually do this, to go ahead and begin the process of planning and preparing for a geoengineering project?"

DAVID KEITH: "Very, very hard to guess. I mean, essential thing to say about this is that technology is the easy part; the *hard part is the politics*. Really deeply hard and almost unguessable. At this point we have no regulatory

Appendix I

structure whatsoever and no treaty structure, so it's really unclear what would—how such a thing would be controlled."

TONY JONES: "Do you have any sort of idea at all what kind of timescale there might be before governments are forced to seriously consider this? Is it 10, 20, 30, 50 years?"

DAVID KEITH: "Well, forced is a very fuzzy word, so a popular thing to say in this business is to say that we would *do it in the case of a climate emergency*. But that's kind of easy to say. In a case of emergency we should do all sorts of wild things, but it's not clear what an emergency is. So I'm a little sticky with the word forced. But I think it could happen any time from a decade from now to many, many decades hence."

The big question right now really is: "*Should we do research in the open atmosphere*? Should we go outside of the laboratory and begin to actually tinker with the system and learn more about whether this will work or not. And I'm somebody who advocates that we do such research."

"And one thing that research may show is that this doesn't work as well as we think. And my view is: whether you're somebody who hopes this will work or hopes it doesn't, more knowledge is a good thing."

TONY JONES: "So if you were given the go-ahead to do research and the funds to do it, because I imagine it would be very expensive, what would you actually do?"

DAVID KEITH: "It's not very expensive actually to begin to do little in-situ experiments. So I am working on one and many other people are. So what we would do – the experiment that I'm most involved with would look at a certain aspect of stratospheric chemistry, of the way that the ozone layer is damaged and we'd be looking at whether or not and how much increase of water vapour in the stratosphere, which may happen naturally, and also the increase of sulphate aerosols if we geoengineered might damage the ozone layer."

"Basically, how much damage there would be and how we could fix it. And that experiment would be done in a very, very small amount of material; we're talking, like, a tonne of material, so small compared to what an aircraft does traveling across the Pacific. And the cost of it would be a "few millions to 5 million" kind of money, which on the scale of big atmospheric research projects is actually not that much. I mean, the total climate research budget is billion class.

TONY JONES: "Is it clear now or is it becoming clearer that the best strategy if you wanted to go to a global scale would be literally flooding the stratosphere with sulphate particles?"

DAVID KEITH: "I think the honest answer has to be that we don't know, that you need to do the research in order to have strong opinions about what's the right answer. I would say, you know, if you really put a gun to my head and said, 'What's the very most likely thing to work right now?' That's probably it. And the reason is because it mimics what nature has done."

"So we have big volcanoes that put sulfur in the stratosphere and we know something about the bad impacts of that and we know something about what it does to cool the planet. And so it seems pretty likely that since we'd be putting in much less than nature puts in, at least for the first half century or more, that we could actually do something and control the risks."

TONY JONES: "Yes, I guess you mentioned volcanic activity and that's what scientists are basing, I suppose, their knowledge on now. What we've seen from volcanic activity is – and you can go back to '91 and *Mount Pinatubo*, which actually caused a fairly sudden drop in global temperatures because it blanketed the atmosphere in that way, but it also had, evidently, climate change effects itself, so there are clearly dangers here."

DAVID KEITH: "For sure. There are a bunch of dangers. There are both the dangers of kind of side effects like ozone loss or interfering with atmospheric chemistry in other ways. There's the basic fact that this is not a perfect compensation for CO_2."

"So for example, carbon dioxide makes the ocean more acidic and doing these things to cool the planet will do nothing to correct that. So in the end we will have to cut emissions no matter what, but the fact that we have to cut emissions in the long run doesn't mean that we might not want to do things in the short run that actually provide real protection, if in fact they do,

Appendix I

protecting people from heat stress or protecting the Arctic from melting."

"So I think we need to get out of the kind of extreme either/or that says you only do this if you can't cut emissions. That's nonsense. Cutting emissions we need to do in order to reduce the risks over the next century or two, but we still might want to do some of this in order to reduce the risks over the next half century and those are really quite distinct things."

TONY JONES: "Let's talk about the risks of actually doing it on a global scale because you've been pretty frank about that. You've actually said you could easily imagine a chain of events that would extinguish life on Earth. Now what would be that potential chain of events from using this kind of technology?"

DAVID KEITH: "Yes, I probably got quoted a little out of context there. I think there are sort of theoretically possible ways that could happen, but I don't think there's socially plausible way it could happen. So, you might in principle be able to put up enough reflective aerosols – probably not sulphates, actually; I think it won't work with sulphates – but some other engineered aerosol."

"And if you did that for 100 years and reflected away sort of 8 per cent of the sunlight, whereas the amount people are talking about doing is more like 1 per cent, then in principle you could actually freeze the oceans over, as happened some good chunk of a billion years ago, and that would be devastating. But I think that the chance of people doing that would sort of be a global suicide is so remote as not to be a serious worry."

"I think the reason I've occasionally said that is that it illustrates the kind of power that this technology grants us. And I think for better, for worse, what this technology gives us is this enormous kind of leverage and power to alter the climate and to do it with a very small amount of money or material and that power should frighten us, I think, and it presents real deep problems for governance."

"If you talk about putting sulphates or some other engineered particle in the stratosphere, the issue is that a very small number of people in principle could do it and have this kind of huge leverage to affect the whole climate in this profound way. And that's what raises the very hard challenge of governance.

TONY JONES: "Yes, is there a fear raised by what you're saying that some country, a superpower, China, for example, has been suggested, could actually do something like this unilaterally and thereby create conflict over the whole idea of geoengineering?"

DAVID KEITH: "Yes, it's certainly possible. So, there's no question it's technically possible to do it unilaterally. So, the actual materials you need, *the aircraft and engineering you need to do this are something that would be in reach easily of any of the G20 states*. It's not hard to do. You could buy the equipment from many aeronautical contractors."

"So in that sense it could be done unilaterally. I think that there are scenarios under which it would happen in the real world unilaterally, but I don't think we should—I mean, I think you can exaggerate that possibility."

"But, you know—so, for example, I think if nothing was done to manage emissions and if climate impacts really fell strongly on, say, India—which might actually happen from heat stress on crops—you could imagine India doing it unilaterally. But there's a kind of a hard and an easy unilateralism."

"So if a country in a really kind of wanton way just starts it with no consultation, that would be clearly ugly, bad, could create conflict, but I think there are also kinds of unilateralism where you're not formally doing it in a legal multinational way, but where you do it with lots of consultation. And in that situation what might happen is a small number of countries might do it and many other countries might publicly say, '*We wish we were involved in the decision*,' and privately say, '*We're pretty happy somebody's doing this because actually it will reduce climate risk and then this other group will take the liability.*' "

TONY JONES: "And final question, because you probably—if someone decided to do this, even if a group of nations decided to do this, there'd be tremendous skepticism in the public and you would, I imagine, get widespread protests, particularly when people realise that with sulphate particles in the atmosphere you'd actually change the colour of the sky, which has a really big psychological effect on people, you would imagine."

Appendix I

"How serious first of all would that change of colour be if you really were able to do it on a global scale and would you expect protests?"

DAVID KEITH: "I think the change of colour would probably be invisible. I think it wouldn't happen. So people have published papers where they get that, but only where they assume a quite large amount of geoengineering. They assume that geoengineering compensates all of the effect of climate change, which I think is a kind of nonsense policy."

"So in a more plausible policy where you gradually wrap this up, compensating only part of the global warming (inaudible), to kind of balance risks and benefits and where you gradually use more advanced particles, maybe starting in 50 years, I think *you never see a change in colour.*"

"So I think that's a bit of a unlikely circumstance. But I do think it's clear that people will protest because there are going to be winners and losers, just as there are under climate change. *So it's important to say that putting CO2 in the atmosphere, which we're doing, creates winners and losers and this will again.*"

TONY JONES: David Keith, we'll have to leave you there. Fascinating to hear from you. We thank you very much for taking the time to come and talk to us.

https://www.chemtrailsplanet.net/2012/11/23/david-keith-promotes-aerosol-geoengineering-as-fix-for-warming-via-covert-climate-modification/

Appendix II
United States Patent and Trademark Office

Geoengineering Related Patents Since 1920

1338343	April 27, 1920: Process and Apparatus for the Production of Intense Artificial Clouds, Fogs, or Mists
1619183	March 1, 1927: Process of Producing Smoke Clouds from Moving Aircraft
1631753	June 7, 1927: Electric Heater, Referenced in 3990987
1665267	April 10, 1928: Process of Producing Artificial Fogs
1892132	December 27, 1932: Atomizing Attachment for Airplane Engine Exhausts
1928963	October 3, 1933: Electrical System and Method
1957075	May 1, 1934: Airplane Spray Equipment
2097581	November 2, 1937: Electric Stream Generator, Referenced in 3990987
2409201	October 15, 1946: Smoke Producing Mixture
2476171	July 18, 1945: Smoke Screen Generator
2480967	September 6, 1949: Aerial Discharge Device
2550324	April 24, 1951: Process for Controlling Weather
2582678	June 15, 1952: Material Disseminating Apparatus for Airplanes
2591988	April 8, 1952: Production of TiO_2 Pigments, Referenced in 3899144
2614083	October 14, 1952: Metal Chloride Screening Smoke Mixture
2633455	March 31, 1953: Smoke Generator
2688069	August 31, 1954: Steam Generator, Referenced in 3990987
2721495	October 25, 1955: Method and Apparatus for Detecting Minute Crystal Forming Particles Suspended in a Gaseous Atmosphere
2730402	January 10, 1956: Controllable Dispersal Device
2801322	July 30, 1957: Decomposition Chamber for Mono-propellant Fuel, Referenced in 3990987
2881335	April 7, 1959: Generation of Electrical Fields
2908442	October 13, 1959: Method for Dispersing Natural Atmospheric Fogs and Clouds
2986360	May 30, 1962: Aerial Insecticide Dusting Device
2963975	December 13, 1960: Cloud Seeding Carbon Dioxide Bullet

3126155	March 24, 1964: Silver Iodide Cloud Seeding Generator, Referenced in 3990987
3127107	March 31, 1964: Generation of Ice-Nucleating Crystals
3131131	April 28, 1964: Electrostatic Mixing in Microbial Conversions
3174150	March 16, 1965: Self-Focusing Antenna System
3234357	February 8, 1966: Electrically Heated Smoke Producing Device
3274035	September 20, 1966: Metallic Composition for Production of Hydroscopic Smoke
3300721	January 24, 1967: Means for Communication Through a Layer of Ionized Gases
3313487	April 11, 1967: Cloud Seeding Apparatus
3338476	August 29, 1967: Heating Device for Use With Aerosol Containers, Referenced in 3990987
3410489	November 12, 1968: Automatically Adjustable Airfoil Spray System with Pump
3429507	February 25, 1969: Rainmaker
3432208	November 7, 1967: Fluidized Particle Dispenser
3441214	April 29, 1969: Method and Apparatus for Seeding Clouds
3445844	May 20, 1969: Trapped Electromagnetic Radiation Communications System
3456880	July 22, 1969: Method of Producing Precipitation From the Atmosphere
3518670	June 30, 1970: Artificial Ion Cloud
3534906	October 20, 1970: Control of Atmospheric Particles
3545677	December 8, 1970: Method of Cloud Seeding
3564253	February 16, 1971: System and Method for Irradiation of Planet Surface Areas
3587966	June 28, 1971: Freezing Nucleation
3601312	August 24, 1971: Methods of Increasing The Likelihood of Precipitation [sic] by the Artificial Introduction of Sea Water Vapor into the Atmosphere Windward of An Air Lift Region
3608810	September 28, 1971: Methods of Treating Atmospheric Conditions
3608820	September 20, 1971: Treatment of Atmospheric Conditions by Intermittent Dispensing of Materials Therein
3613992	October 19, 1971: Weather Modification Method
3630950	December 28, 1971: Combustible Compositions for Generating

Appendix II

	Aerosols, Particularly Suitable for Cloud Modification and Weather Control and Aerosol-ization Process
USRE29142	This patent is a reissue of patent US3630950, Combustible compositions for generating aerosols, particularly suitable for cloud modification and weather control and aerosol-ization process
3659785	December 8, 1971: Weather Modification Utilizing Micro-encapsulated Material
3666176	March 3, 1972: Solar Temperature Inversion Device
3677840	July 18, 1972: Pyrotechnics Comprising Oxide of Silver for Weather Modification Use
3722183	March 27, 1973: Device for Clearing Impurities From the Atmosphere
3769107	October 30, 1973: Pyrotechnic Composition for Generating Lead Based Smoke
3784099	January 8, 1974: Air Pollution Control Method
3785557	January 15, 1974: Cloud Seeding System
3795626	March 5, 1974: Weather Modification Process
3808595	April 30, 1974: Chaff Dispensing System
3813875	June 4, 1974: Rocket Having Barium Release System to Create Ion Clouds In The Upper Atmosphere [*sic*]
3835059	September 10, 1974: Methods of Generating Ice Nuclei Smoke Particles for Weather Modification and Apparatus Therefore
3835293	September 10, 1974: Electrical Heating Apparatus [*sic*] for Generating Super Heated Vapors
3877642	April 15, 1975: Freezing Nucleant
3882393	May 6, 1975: Communications System Utilizing Modulation of the Characteristic Polarization of The Ionosphere
3896993	July 29, 1975: Process for Local Modification of Fog and Clouds for Triggering Their Precipitation and for Hindering the Development of Hail Producing Clouds
3899129	August 12, 1975: Apparatus for generating ice nuclei smoke particles for weather modification
3899144	August 12, 1975: Powder contrail generation
3940059	February 24, 1976: Method for Fog Dispersion
3940060	February 24, 1976: Vortex Ring Generator

3990987	November 9, 1976: Smoke generator
3992628	November 16, 1976: Countermeasure system for laser radiation
3994437	November 30, 1976: Broadcast dissemination of trace quantities of biologically active chemicals
4042196	August 16, 1977: Method and apparatus for triggering a substantial change in earth characteristics and measuring earth changes
RE29,142	February 22, 1977: Reissue of: 03630950, Combustible compositions for generating aerosols, particularly suitable for cloud modification and weather control and aerosol-ization process
4035726	July 12, 1977: Method of controlling and or improving high-latitude and other communications or radio wave surveillance systems by partial control of radio wave etc.
4096005	June 20, 1978: Pyrotechnic Cloud Seeding Composition
4129252	December 12, 1978: Method and apparatus for production of seeding materials
4141274	February 27, 1979: Weather modification automatic cartridge dispenser
4167008	September 4, 1979: Fluid bed chaff dispenser
4347284	August 31, 1982: White cover sheet material capable of reflecting ultraviolet rays
4362271	December 7, 1982: Procedure for the artificial modification of atmospheric precipitation as well as compounds with a dimethyl sulfoxide base for use in carrying out said procedure
4402480	September 6, 1983: Atmosphere modification satellite
4412654	November 1, 1983: Laminar micro-jet atomizer and method of aerial spraying of liquids
4415265	November 15, 1983: Method and apparatus for aerosol particle absorption spectroscopy
4470544	September 11, 1984: Method of and Means for weather modification
4475927	October 9, 1984: Bipolar Fog Abatement System
4600147	July 15, 1986: Liquid propane generator for cloud seeding apparatus
4633714	January 6, 1987: Aerosol particle charge and size analyzer
4643355	February 17, 1987: Method and apparatus for modification of climatic conditions
4653690	March 31, 1987: Method of producing cumulus clouds

Appendix II

4684063	August 4, 1987: Particulates generation and removal
4686605	August 11, 1987: Method and apparatus for altering a region in the earth's atmosphere, ionosphere, and or magnetosphere
4704942	November 10, 1987: Charged aerosol
4712155	December 8, 1987: Method and apparatus for creating an artificial electron cyclotron heating region of plasma
4744919	May 17, 1988: Method of dispersing particulate aerosol tracer
4766725	August 30, 1988: Method of suppressing formation of contrails and solution therefor
4829838	May 16, 1989: Method and apparatus for the measurement of the size of particles entrained in a gas
4836086	June 6, 1989: Apparatus and method for the mixing and diffusion of warm and cold air for dissolving fog
4873928	October 17, 1989: Nuclear-sized explosions without radiation
4948257	August 14, 1990: Laser optical measuring device and method for stabilizing fringe pattern spacing
1338343	August 14, 1990: Process and Apparatus for the production of intense artificial Fog
4999637	March 12, 1991: Creation of artificial ionization clouds above earth
5003186	March 26, 1991: Stratospheric Welsbach seeding for reduction of global warming
5005355	April 9, 1991: Method of suppressing formation of contrails and solution therefor
5038664	August 13, 1991: Method for producing a shell of relativistic particles at an altitude above the earths [sic] surface
5041760	August 20, 1991: Method and apparatus for generating and utilizing a compound plasma configuration
5041834	August 20, 1991: Artificial ionospheric mirror composed of a plasma layer which can be tilted
5056357	October 15, 1991- Acoustic method for measuring properties of a mobile medium
5059909	October 22, 1991: Determination of particle size and electrical charge
5104069	April 14, 1992: Apparatus and method for ejecting matter from an aircraft

5110502	May 5, 1992: Method of suppressing formation of contrails and solution therefor
5156802	October 20, 1992: Inspection of fuel particles with acoustics
5174498	December 29, 1992: Cloud Seeding
5148173	September 15, 1992: Millimeter wave screening cloud and method
5245290	September 14, 1993: Device for determining the size and charge of colloidal particles by measuring electro-acoustic effect
5286979	February 15, 1994: Process for absorbing ultraviolet radiation using dispersed melanin
5296910	March 22, 1994: Method and apparatus for particle analysis
5327222	July 5, 1994: Displacement information detecting apparatus
5357865	October 25, 1994: Method of cloud seeding
5360162	November 1, 1994: Method and composition for precipitation of atmospheric water
5383024	January 17, 1995: Optical wet steam monitor
5425413	June 20, 1995: Method to hinder the formation and to break-up overhead atmospheric inversions, enhance ground level air circulation and improve urban air quality
5434667	July 18, 1995: Characterization of particles by modulated dynamic light scattering
5441200	August 15, 1995: Tropical cyclone disruption
5486900	January 23, 1996: Measuring device for amount of charge of toner and image forming apparatus having the measuring device
5556029	September 17, 1996: Method of hydrometer dissipation (clouds)
5628455	May 13, 1997: Method and apparatus for modification of supercooled fog
5631414	May 20, 1997: Method and device for remote diagnostics of ocean-atmosphere system state
5639441	June 17, 1997: Methods for fine particle formation
5762298	June 9, 1998: Use of artificial satellites in earth orbits adaptively to modify the effect that solar radiation would otherwise have on Earth's weather
5912396	June 15, 1999: System and method for remediation of selected atmospheric conditions
5922976	July 13, 1999: Method of measuring aerosol particles using automated mobility-classified aerosol detector

Appendix II

5949001   September 7, 1999: Method for aerodynamic particle size analysis
5984239   November 16, 1999: Weather modification by artificial satellite
6025402   February 15, 2000: Chemical composition for effectuating a reduction of visibility obscuration, and a detoxification of fumes and chemical fogs in spaces of fire origin
6030506   February 29, 2000: Preparation of independently generated highly reactive chemical species
6034073   March 7, 2000: Solvent detergent emulsions having antiviral activity
6045089   April 4, 2000: Solar-powered airplane
6056203   May 2, 2000: Method and apparatus for modifying supercooled clouds
6110590   August 29, 2000: Synthetically spun silk nano-fibers and a process for making the same
6263744   July 24, 2001: Automated mobility-classified-aerosol detector
6281972   August 28, 2001: Method and apparatus for measuring particle-size distribution
6315213   November 13, 2001: Method of modifying weather
6382526 fibers   May 7, 2002: Process and apparatus for the production of nano-fibers
6408704 apparatus   June 25, 2002: Aerodynamic particle size analysis method and apparatus
6412416   July 2, 2002: Propellant-based aerosol generation devices and method
6520425   February 18, 2003: Process and apparatus for the production of nano-fibers
6539812   April 1, 2003: System for measuring the flow-rate of a gas by means of ultrasound
6553849   April 29, 2003: Electrodynamic particle size analyzer
6569393   May 27, 2003: Method and device for cleaning the atmosphere

An Extensive List of Patents on weather modification 1920–Oct 5 2012
http://www.geoengineeringwatch.org/links-to-geoengineering-patents/

Websites and Resources
Websites:

www.geoengineeringwatch.org
www.carnicominstitute.org
Nature Bats Last, *www.guymcpherson.com*
Arctic Methane Group, *www.ameg.me*
www.tabublog.com
www.skyderalert.com
www.toxicsky.org
www.weatherwars.info
www.aircrap.org
www.agriculturedefensecoalition.org
www.climateviewer.com

Books:

Policy Implications of Greenhouse Warming, National Academy Press, 1992.
Chemtrails, HAARP, and the Full Spectrum Dominance of Planet Earth, Elana Freeland, 2014.
Going Dark, Guy R. McPherson, 2013.
Red Sky at Morning, James G. Speth, 2004.
Dancing Under the Red Star, Karl Tobien, 2006.
Peak Everything, Richard Heinberg, 2007.
Tragedy & Hope, Carol Quigley, 1966.
In the Absence of the Sacred, Jerry Mander, 1991.
The Secret History of American Empire, John Perkins, 2007.
Dissolving Illusions; Disease, Vaccines and the Forgotten History, Dr. Suzanne Humphries and Roman Bystrianyk, 2013.
Murder by Injection; The Story of the Medical Conspiracy Against America, Eustace Mullins, 1988.
Ecopsychology; Restoring the Earth, Healing the Mind , Theodore Roszak, Mary Gomes, and Allen D. Kanner, 1995.
The Ecology of Commerce, Paul Hawkens, 1993.

Links:

Recommended National Program In Weather Modification, 1966; Associate Administrator for Space Science Application, NASA.
http://www.geoengineeringwatch.org/documents/19680002906_1968002906.pdf

NASA Future Strategic Issues/Future Warfare. [*c. 2025*]
https://tabublog.com/2016/01/20/nasa-is-a-government-military-operation/

Documentation of Geoengineering Practices
http://www.geoengineeringwatch.org/documents-2/

Rights of Nature Law
https://tabublog.com/2014/11/05/mendocino-county-ca-makes-history-and-passes-law-establishing-local-self-governance/

Health and Detox
www.DWDGshop.com
www.CTbusters.com

James (Jamie) Lee resides near the Mendocino Coast of Northern California growing biodynamic and organic food at his 100 year old farm. He graduated from business school at San Diego State University and attended the GreenMBA program at New College in Santa Rosa, California.

He has had over 25 years experience working for the investment banking firm, Furman, Selz., Inc. in New York City and Robertson, Coleman, Stephens in San Francisco, before starting his own investment research firm in 1991.

Jamie has used his Wall Street experience for the past 10 years, 6–10 hours a day to explore in depth, investigate, and understand the occulted, hidden world of finance, government, and secret societies that run all that very few people are aware of, their existance.

You can find his research on many important and relevant topics on the following websites:

www.tabublog.com
www.aplanetruth.info
www.AVVI.info

Visit his YouTube channel: *Aplanetruth.info*

www.ingramcontent.com/pod-product-compliance
Lightning Source LLC
Chambersburg PA
CBHW060353190526
45169CB00002B/580